Goodman Fielder

As Australia's largest consumer of wheat and edible oils, Goodman Fielder is proud to be a leading participant in *Natural Quality: The Produce of Australia*.

While most people are familiar with our retail brand names such as Uncle Tobys, Meadow Lea, Buttercup and White Wings, many people do not know that we are Australasia's largest and most diversified food company.

We have businesses as diverse as milling and baking, margarine and cooking oils, and gelatine and meal solutions, and customers that range from supermarkets to restaurants to industrial clients.

We are continually improving our businesses and we remain focused on our vision of becoming a world-class Australasian-based food company generating sustainable, superior returns for all shareholders.

We operate in an environment that is increasingly global and competitive and we must continue to respond to the major consumer trends facing food companies across the world – quality assurance and food safety, consumer demand for convenience and value, and the desire for nutritious foods.

We are already responding to these trends by targeting the fast-growing home meal replacement market, launching increasingly nutritious, value-added products and introducing strict quality assurance programs.

As a demonstration of our commitment to quality assurance and food safety, we have developed the Goodman Fielder Trust Mark, which will eventually appear on the consumer panel of every one of our products.

The Goodman Fielder Trust Mark supports our promise of 'Good Food Guaranteed' and is a tangible sign to our customers that we are committed to producing food of the highest quality – nutritious, good-tasting and safe.

Natural Quality: The Produce of Australia presents a powerful message to the international market: that Australia is a major producer and supplier of the safest and highest quality food in the world.

Goodman Fielder is proud to be one of those suppliers.

Goodman Fielder Limited

NATURAL QUALITY

THE PRODUCE OF AUSTRALIA

FOCUS

Focus Publishing Pty Ltd

Enquiries should be addressed to the publisher:

Focus Publishing Pty Ltd
ACN 003 600 360
PO Box 518, Edgecliff NSW 2027
Telephone: (02) 9327 4777
Fax: (02) 9362 3753
Email: focus@focus.com.au
Internet: //www.focus.com.au/

Chairman: Steven Rich
Publisher: Jaqui Lane
Associate Publisher: Jennifer Walkden
Publishing Manager: Rod de Martin
Production Manager: Timothy Ho
Editor: Paul Chan
Designer: Penny Ruthberg
Client Services Co-ordinator: Tanya Simpson
Photo Researcher: Karen Burgess

ISBN 1 875359 53 2

ABOUT THE PUBLISHERS

Focus Publishing is a leading corporate publishing house based in Sydney, Australia.

We provide an integrated service in the creation, production, funding and distribution of books for major corporates, government, business and industry associations, and educational institutions.

Focus has published over 60 titles for many of Australia's leading companies and organisations and has established an enviable reputation for high quality and performance.

In a commercial environment, books and high-quality publications are formidable marketing tools. Focus has the proven expertise to bring together creative talent and corporate marketing objectives to produce publications that have real utility.

Our clients include Qantas, Woolworths, Colonial, the Commonwealth Bank, Westfield, the Snowy Mountains Authority and most State Governments.

CONTENTS

Advisory Committee 4
Roll of Honour 5
Foreword 6
Introduction 7

CHAPTER ONE
The Australian Agri-Food Industry – An Overview 8
Participants: Goodman Fielder, State Government of Victoria, National Farmers' Federation, Agriculture, Fisheries and Forestry – Australia, IHD Pty Ltd

CHAPTER TWO
Technology and Infrastructure 34
Participants: State Government of Victoria, P&O Cold Storage

CHAPTER THREE
A Safe and Superior Product – Pure Quality Assurance 42
Participants: Rangers Valley Cattle Station Pty Ltd, Food Safety Victoria, Darling Downs Bacon Co-operative Association Ltd

CHAPTER FOUR
Sustainable Solutions 50
Participants: State Government of Victoria, Business Ballarat

CHAPTER FIVE
A World-Class Supplier 58
Participants: Maxi-Cube, Cargill Australia Ltd, Ridley Corporation Ltd, Wirrilla Pty Ltd

CHAPTER SIX
Meat 74
Participants: Darling Downs Bacon Co-operative Association Ltd, Columbus Line Australia Pty Ltd, Glenmore Export Australia, Stanbroke Pastoral Company Pty Ltd

CHAPTER SEVEN
Dairy and Eggs 90
Participants: Warrnambool Cheese & Butter Factory Company Ltd, Pace Farms Pty Ltd

CHAPTER EIGHT
Seafood 100
Participants: Nortas Pty Ltd, Tassal Ltd

CHAPTER NINE
Cereals 110
Participants: AWB Limited, The Lentil Company Pty Ltd

CHAPTER TEN
Fruit and Vegetables 120
Participants: Australian Pure Fruits Pty Ltd, Asparagus Growers of Sunraysia Inc, Costa's Pty Ltd, Moraitis Fresh Packaging (NSW) Pty Ltd, Ardmona Fresh Fruit Packers Pty Ltd, edp Packaging and Handling Systems, Gartner Farms, Melbourne Markets, Harvest Moon, Perfecta Produce, Woolworths Limited

CHAPTER ELEVEN
Wines and Beverages 144
Participants: Berri Ltd, Bundaberg Brewed Drinks Pty Ltd, Roberts Estate Wines, Gartner Wines, The Original Juice Company, Orlando Wyndham Group Pty Ltd

CHAPTER TWELVE
Processed Foods 160
Participants: Ricegrowers' Co-operative Limited, Unilever Foods

CHAPTER THIRTEEN
Oils and Specialty Foods 174
Participants: Gardner Smith Pty Ltd, Australian Food Merchants, Chateau Yaldara Pty Ltd, Olivegrove Trading Co Pty Ltd

Directory 184
Internet Information 206
Picture Credits 207
Writers' Credits 208

ADVISORY COMMITTEE

CHAIRMAN

Reg Clairs, Former Group Managing Director and Chief Executive Officer, Woolworths Limited

COMMITTEE

Enzo Allara, Chairman, Unifoods
John Baker, Chief Executive, Produce Marketing Australia
John Cox, Managing Director, Stanbroke Pastoral Co Pty Ltd
Dr Wendy Craik, Executive Director, National Farmers' Federation
Julian Cribb, Director, CSIRO National Awareness, CSIRO Australia
Ian Donges, President, National Farmers' Federation
Achim Drescher, Managing Director, Columbus Line Australia Pty Ltd
David Farley, Managing Director, Colly Cotton Limited
Kevin Hayes, Chairman and Managing Director, Cadbury Schweppes Australia Limited
Dr John Keniry, Chairman, Ridley Corporation Limited
Ted Kunkel, President and Chief Executive Officer, Foster's Brewing Group Limited
David Mackay, Managing Director, Kellogg (Aust.) Pty Ltd
Donald McGauchie, Council Member, National Farmers' Federation and Supermarket to Asia
Fay McGuigan, Export Director, Brian McGuigan Wines Limited
Jean McRuvie, Chief Executive Officer, Cattlemen's Union of Australia Inc.
Nicholas Moraitis, Chairman and Chief Executive Officer, Moraitis Holdings Pty Ltd
Mark Napper, Managing Director, Australian Horticultural Corporation
John Penn, Marketing Director, Superfood Ideas
John Randall, Managing Director, P&O Cold Storage Ltd
Steven Rich, Chairman, Focus Publishing
Chris Roberts, Managing Director, Arnotts Limited
Rob Robson, Managing Director, Harvest Fresh Cuts Pty Ltd
Murray Rogers AM, Chief Executive Officer, AWB Limited
Brian Scarsbrick, Chief Executive, Landcare Australia Limited
Doug Shears, Executive Chairman, ICM Australia Pty Limited
Peter Shelley, Managing Director, Tassal Limited
Dr Bruce Standen, Advisor, Meat and Livestock Australia
Alan Tooth, Managing Director, Dairy Farmers Group
Ruth Wade, Director, Meat and Livestock Australia

ROLL OF HONOUR

LEAD PARTNERS

Goodman Fielder Limited

Food Victoria

Business Victoria

Department of Natural Resources and Environment

Food Safety Victoria

Melbourne Market Authority

MAJOR PARTNERS

Agriculture, Fisheries and Forestry – Australia (AFFA)

Unilever Foods (a division of Unilever Australia Ltd)

KEY PARTNERS

Cargill Australia Limited

Columbus Line Australia Pty Ltd

Costa's Pty Ltd

Darling Downs Bacon Co-operative Association Ltd

IHD Pty Ltd

Moraitis Fresh Packaging (NSW) Pty Ltd

CONTRIBUTING PARTNERS

Ardmona Fresh Fruit Packers Pty Ltd
AWB Limited
Berri Ltd
Bundaberg Brewed Drinks Pty Ltd
edp Packaging and Handling Systems
Gardner Smith Pty Ltd
Gartner Farms
Gartner Wines
Glenmore Export Australia Pty Ltd
Harvest Moon
MaxiTRANS Industries Limited
Pace Farms Pty Ltd
Perfecta Produce
Rangers Valley Cattle Station Pty Ltd
Ricegrowers' Co-operative Limited
Ridley Corporation Limited
Roberts Estate Wines
The Lentil Company Pty Ltd
Warrnambool Cheese & Butter Factory Company Limited
Wirrilla Pty Ltd
Woolworths Limited

AFFILIATE PARTNERS

Asparagus Growers of Sunraysia Inc
Australian Food Merchants
Australian Pure Fruits Pty Ltd
Business Ballarat
Chateau Yaldara Pty Ltd
National Farmers' Federation
Nortas Pty Ltd
Olivegrove Trading Co Pty Ltd
Orlando Wyndham Group Pty Ltd
P & O Cold Storage
Stanbroke Pastoral Company Pty Ltd
Tassal Limited
The Original Juice Company

FOREWORD

It is with great pleasure that I endorse *Natural Quality: The Produce of Australia* as a major contribution to highlighting Australia's capacity as a supplier of a wide range of safe, high-quality foods produced in a clean environment.

Australia is fortunate to enjoy clean air and water, reflecting our smaller population, the absence of significant trans-border pollution, and our location in the less industrialised Southern Hemisphere. As an island nation we also remain free of many of the animal and plant diseases found elsewhere in the world, and we give a high priority to enforcing our quarantine controls.

In addition to these natural competitive advantages, the Australian agri-food industry is actively embracing production and processing technology and quality assurance systems. These are enhancing Australia's 'clean and green' credentials and help to ensure that our customers can continue to have confidence in the wholesomeness, safety and quality of our products. The widespread and rapid adoption of technical innovations is facilitated by our world-class research and development infrastructure, which is strongly supported by the Australian Government.

Australia's ability to continue to deliver safe, high-quality food is also underpinned by the important role that government has in establishing and enforcing comprehensive food standards regulation and monitoring to ensure compliance. Crucial government functions also include quarantine, export inspection, exotic pest and disease contingency planning, natural resource management and assisting the agri-food industries to adopt quality assurance systems.

The Australian Government's commitment to the Australian food industry is demonstrated by the establishment of the Prime Minister's Supermarket to Asia Council. The Council, chaired by the Prime Minister, is bringing together government and industry leaders to provide strategic direction and remove any impediments to Australia continuing to be a world leader in food exports. A particular focus of the Council has been to work with industry on a through-chain basis to continually improve quality and safety performance.

Australia's food companies recognise that they cannot afford to become complacent and simply rest on the successes of the past. I am confident that they will continue to produce and deliver the varied and high-quality products demanded by today's consumers.

The organisations profiled in this book will play a large part in this absorbing and challenging task, and the Australian Government is proud to support their efforts.

Mark Vaile
Minister for Agriculture, Fisheries and Forestry

INTRODUCTION

At the national Outlook 1996 conference, I articulated a vision for the Australian agri-food industry, based on the ideal that we should never be satisfied with established methods of doing things – rather, we should seek to create our future as a nation, now.

That vision is that by the year 2020, Australia will become a significant global supplier of food. By this, I mean the kind of quality products we are capable of producing.

The world is beginning the biggest surge of food demand in its history and Australia is perfectly poised to ride the crest of this wave.

In order to take advantage of this potential, we must ensure that our food exports meet or exceed the quality expectations of the consumers. The safety of the food is non-negotiable and must fall below the standard minimum chemical residues, be hygienically packaged and be competitive in all respects with imports from overseas countries.

Australia is blessed with one of the cleanest environments in the world, encompassing an abundance of land and sea, different climatic zones suitable for growing a wide variety of produce, and a precious network of rivers irrigating fertile agricultural regions. Supplementing these natural advantages are a considerable body of knowledge, an advanced research and technology infrastructure and a talented workforce.

Australia's talented migrant population has introduced exciting new flavours into the nation's diet and injected new know-how and vigour into the agri-food industry. Australian agribusinesses now have valuable first-hand experience of catering for the tastes of different cultures before they even begin exporting.

Our close proximity to Asia means that we are well placed to tap into one of the fastest growing food markets in the world. The Australian Government has provided major assistance to Australian companies keen to explore this market by establishing the Supermarket to Asia Council.

Three years down the track from that original vision statement, I am pleased to see that Australia is firmly on track to meeting that vision. To mark how far we have come, I commend to you *Natural Quality: The Produce of Australia.*

This book is an authoritative review of the achievements of Australia's agribusiness sector and a beautifully illustrated, inspiring statement of how much further we can go.

I take this opportunity to congratulate the partner companies and government bodies herein and to thank them for sharing the vision for our country's agri-food industry.

Reg Clairs
Former Group Managing Director and Chief Executive Officer
Woolworths Limited

CHAPTER ONE

The Australian Agri-food Industry

AUSTRALIAN AGRIBUSINESS COMES OF AGE

Australia's multi-billion-dollar agri-food industry is tapping into a wave of international consumer awareness of the environment and food production to position itself for growth into the next millennium.

The industry is in a state of metamorphosis, with innovators in all sectors successfully working to enhance the nation's reputation as an exporter of high-quality, wholesome agri-foods.

There is now a real urgency within the industry, as traditional global trade barriers are dismantled and the race begins in earnest to win and secure market share.

Australia, fortunately, is well placed to capitalise on the forces of globalisation, with many agricultural industries having been exposed to the winds of free trade long before many of their international counterparts were compelled to do so under legislation applied by the World Trade Organisation.

From humble beginnings and an agricultural system modelled on the nation's Anglo-Saxon heritage, the food industry has evolved into a diverse, multicultural system which supports more than 130,000 farmers and 4,000 food manufacturers, who employ about 170,000 workers.

Turnover in the processed food and beverage industries alone has more than doubled in the past 10 years, from $16 billion to around $40 billion per year.

The agri-food industry is the nation's largest manufacturing sector, accounting for about 20 per cent of total manufacturing turnover.

Above: *Australia's agri-food industry is its largest manufacturing sector, accounting for about 20 per cent of total manufacturing turnover. Having been exposed to free trade long before many of their international counterparts, Australian agribusinesses are well placed to capitalise on the forces of globalisation*

VALUE ADDING IS THE KEY

Exports of value-added food are a driving force behind the development of the agri-food industry. This industry has remained a strong export performer in the face of regional economic fluctuations.

Processed foods are crucial to Australia's standing as an exporting nation, earning the nation more than $11 billion per year and accounting for 15 per cent of total exports.

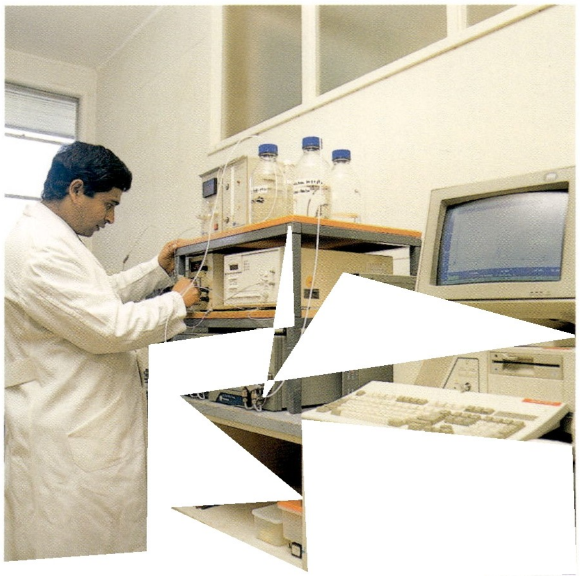

There has been a steady rise in the past decade in the number of firms becoming involved in highly value-added production.

Analysis by the Australian Bureau of Statistics shows that during the first half of the decade, most industries recorded strong growth in terms of value adding. The sectors with the most significant gains were flour and cereal foods, dairy products, beverages and malt.

Exports of highly processed products are rapidly becoming a feature of the nation's agribusiness industry. By 1995–96, the value of this sector had jumped by 128 per cent to more than $4.5 billion.

The majority of growth occurred in the Japanese marketplace, which is Australia's largest market. Exports of Australian food to Japan have averaged more than $3 billion per year.

The country's growth as a market for highly value-added products consistently outstrips the demand for minimally processed food and beverages.

With Australia's workforce increasingly becoming skilled at operating in both Asian and Western environments, the expansion into other emerging Asian markets is a logical progression.

Ten of Australia's 14 largest export markets are in the Asian region. As a result of rising living standards and a growing middle class in the region, it is expected that more of these countries will soon present similar opportunities for Australian agribusiness.

DEVELOPMENT IN MAJOR MARKETS

Industrial development in Asia continues to generate synergies between Australia's resource-rich economy and the relatively resource-poor economies of north Asia.

To date, growth in food exports to these regions has been in the unprocessed and minimally processed sectors, but long-term trends suggest strong, emerging market opportunities for highly processed Australian food and beverages.

The embracing of change has been the hallmark of Australian agriculture's 'coming of age'.

Top and above: *An uncompromising approach to food safety and quality assurance is a vital ingredient in the success of Australian agribusiness. With the adoption of systematic regimes such as Hazard Analysis and Critical Control Points, the agri-food industry is ensuring that it maintains one of the most stringent quality frameworks in the world*

Opposite: *An aerial view of sugarcane fields in northern Queensland*

Overleaf: *Machine-harvesting wine grapes in New South Wales' Hunter Valley. Australia's wine industry is one of its many international agribusiness success stories, with Australian wines selling well throughout the world and regularly winning international awards*

MULTICULTURAL SOCIETY ADDS FLAVOUR

The growth of a truly multicultural society in Australia has infused the agri-food industry with the cultures, tastes and talents of people from many nations.

Initially it was the Greeks, Italians, Belgians, Czechs and Turks who broadened the scope of those cultivating the nation's food bowl, but lately, it has been the Vietnamese, Thai, Lebanese and Chinese who have further contributed to the diversification process.

74

Top: *A cattle sale in full swing at the Colac saleyards in Victoria. The meat industry is a staple of Australian agribusiness and a multi-billion-dollar income earner*

Above: *A cut of high-quality beef produced by Rangers Valley Cattle Station in New South Wales. The cattle are fed a ration of corn mixed with other quality ingredients to produce fine-tasting beef for the demanding Japanese market*

Opposite: *A stockman musters Angus cattle along a spur in Victoria's Otway Ranges*

This diversity makes Australia the ideal proving ground for the development of products, techniques and services aimed for both domestic and international markets.

The nation's multicultural base is also proving to be a vital ingredient in moves by Australian firms to target import replacement opportunities.

Food production industries, such as the cultivation of Chinese vegetables – once unheard of in Australia – are now growing at a faster rate than many of the more traditional sectors.

The realisation that the future of the Australian food industry is inextricably linked with engaging the Asian region is gaining momentum.

This is a trend that is likely to continue unabated, in spite of currency fluctuations within the region.

EMBRACING OPPORTUNITY

As a direct result of industry deregulation, Australian agribusiness is more responsive than ever to the development of new avenues of opportunity.

The average effective rate of assistance for food and beverage manufacturing has been estimated at just three per cent, compared to nine per cent for manufacturing as a whole.

There is a realisation that research and development (R & D) is crucial in providing Australian agribusiness with a strategic advantage.

Food industry spending on R & D amounts to about $130 million per year, or about eight per cent of the total amount spent on R & D in the manufacturing sector. More than 95 per cent of food industry R & D expenditure is funded by individual companies.

Investment in Australian agribusiness continues to grow. Private new capital expenditure is increasing by about ten per cent per year.

In terms of profitability, the industry's performance is also strong. For the ten years to 1995–96, food company profits before tax increased by 118 per cent, compared with a rise of 102 per cent for total manufacturing.

Australian firms have a well-earned reputation for growth through innovation and for being responsive to the needs of other countries.

INGENUITY AND PERSISTENCE

There are numerous examples of Australian ingenuity and persistence at work.

Unilever Foods – a division of Unilever Australia – is utilising premium-quality produce from the nation's most productive agricultural land to forge new markets in the Pacific rim.

A stringent in-house quality and safety regime often exceeds international standards, and the firm is a leader in the development of the Hazard Analysis Critical Control Point (HACCP) system of food safety.

Unilever Foods is also proving to be a champion of the environmental movement, through its work in pioneering avenues of waste reduction, energy conservation and recycling.

By utilising a system already implemented at Unilever's Merced tomato processing plant in California, a large portion of the nutrient-rich waste water from the firm's Tatura plant, in the heart of Victoria's Goulburn Valley food bowl, is now being recycled and utilised in irrigation as fertiliser.

Goodman Fielder, the rising star of the Australian food industry, is a company that is maximising the potential of the agribusiness sector.

The company spends more than one per cent of its substantial annual turnover on R & D and has more than 25 R & D centres around the world.

These centres, such the Ingredients Centre at Botany, are also accessible to Goodman Fielders' many customers. Such pro-active alliances are streamlining the process of R & D delivery, particularly in the field of product development.

REWARDS FOR BUILDING ALLIANCES

The success of Australia's agri-food sector bears testament to an industry which has worked solidly during the last decade to improve its information and trade linkages.

These bonds have allowed Australian farmers and manufacturers to develop, through the sharing of knowledge and technology, closer alliances which, in turn, are leading to greater efficiency.

Collaborative research projects, a two-way flow of information via the Internet and industry forums, and a greater level of feedback from consumers at home and abroad are reshaping the food industry.

Above: *Racks of strawberries grown using the hydroponic method in Queensland. Australian agribusinesses are quick to embrace new technology, realising that research and development are critical to the success of the industry*

Opposite (top): *A view of cultivated fields in New South Wales. With a stabilising population but an ever-increasing agricultural output, Australia is well placed to supply the food requirements of the world's growing population*

Opposite (bottom): *A box of crisp, juicy Australian apples destined for foreign shores*

Food Victoria – an initiative of the Victorian Government and the food industry – is one such forum which encompasses many of the aims of Australian agribusiness as it heads into the 21st century.

Targets of $6 billion in food exports from Victoria by 2001 and $12 billion by 2010 have been set, and initiatives have been put in place to allow the Victorian agri-food industry to lift its exports from the current $4 billion per year.

Central to the program are improved linkages between all sectors via the increased sharing of knowledge.

Food Victoria is working to establish on-line opportunities which will allow readers to tap into information on food legislation, food suppliers and distributors, regional investment opportunities, educational services, and State and national assistance programs.

In promoting its role to help reshape Australian agribusiness, Food Victoria points to the fact that Internet trading must be utilised if the industry, as a whole, is to grow.

OUTWARD VIEW FOR EXPORTING SUCCESS

Australia's population is stabilising, with growth rates of less than 1.8 per cent per year, and increasingly, the domestic market will be incapable of sustaining the food industry's expansion.

However, with scientists predicting that, due to global population growth, the world will have to grow more food in the next three decades than during the whole of human history to date, Australia's position as a premium food provider seems to be assured.

As Australia's agribusiness industry positions itself for the new millennium, it must work methodically through short-term challenges in order to secure long-term gains.

David Gorrie

Goodman Fielder

Clockwise from above: *Goodman Fielder's portfolio of powerful brands and products is one of its major strengths. Goodman Fielder is number one or two in virtually every market in which it operates; wheat harvesting at Horsham, Victoria; canola field, Wagga Wagga, New South Wales; Uncle Toby's Vita Gold bread*

Seeing another of its products go onto supermarket shelves, working with a chocolate manufacturer to put more texture into its mousse, purchasing the licence to produce pasta sauce and Indian meals in Australia – it's all in a day's work for Goodman Fielder.

Australia's largest food manufacturer, with operations in more than 40 countries, the company is the market leader with many of its products and owns the leading brands in virtually all its businesses. These encompass the production and delivery of a broad range of consumer foods and ingredients for the retail, food service, commercial and industrial sectors. Product types, of which there are over 32, range from cake to pasta, cereal to bread, and cooking oil to gelatin.

Ranked today by *Fortune* magazine as one of the world's top 50 food producers, Goodman Fielder's beginnings were a little more modest. Geo Fielder was its founder, setting up a milling company in 1909 and establishing a heritage that has been built on to the present. In 1986, an acquisition saw the formation of Goodman Fielder Limited, and in the following 10 years, further mergers brought together six of Australia's and New Zealand's leading food companies, creating Goodman Fielder as it exists today.

Over the years, judicious acquisitions and divestments have continued, helping the company grow and develop its vision of becoming a truly world-class Australasian-based food business.

In one year alone Goodman Fielder bought New Zealand-based Defiance Mills, the Bunge milling and baking business and the Burns Philp Consumer Foods Business, including local licences to produce the Paul Newman range of sauces and dressings in Australia and Patak's Indian products. In January 1998, Goodman Fielder finalised the acquisition of the edible gelatin operations of United States company Hormel, incorporating it into Goodman Fielder Ingredients – a move that makes Goodman Fielder the largest supplier of edible gelatin products in the world.

It is for manufacturing some very famous products that Goodman Fielder is best known. Uncle Toby's cereals, White Wings cake mixes, Buttercup bread and Meadow Lea margarine have become Australian icons, with an appeal that crosses the generations.

Goodman Fielder's food production is divided into two groups: consumer and commercial, to provide greater customer focus. Meadow Lea margarine has been the best-selling brand in Australia for 20 years, Uncle Toby's breakfast cereals are number two, and

Uncle Toby's muesli bars are number one in their respective markets. Cooking oils and salad dressings are among the most popular supermarket choices.

While Goodman Fielder is in the enviable position of owning many of the leading brands in its businesses, be they spreads, snacks or cereals, the company's success has not made it complacent. The food production business is constantly evolving, and Goodman Fielder is responding quickly to consumer lifestyle changes. Meadow Lea now comes in salt-reduced and low-fat versions and Uncle Toby's has released many new products that meet modern demands for fat-reduced foods, such as Lite Start, a folate- and Vitamin E-enriched, low-fat breakfast cereal. Folate, an important nutrient for women of child-bearing age, is also present in Uncle Toby's VitaGold bread, with two slices providing over 50 per cent of the recommended dietary intake. The company's chicken fillets and portions, with the Heart Foundation's 'tick of approval', contain as little as one per cent fat.

Goodman Fielder's motto – 'Good Food Guaranteed' – is the unifying force behind the company's many businesses. These words assure customers that Goodman Fielder is committed to nothing but the best, by world standards. To the company, living up to this vision

Clockwise from above: *Confectionery pilot plant at the Ingredients Research facility at Botany, New South Wales; Goodman Fielder's range of healthy, tasty products are enjoyed by children of all ages; Allied Flour and Baking Services manufacture a range of quality flours and meals for the baking and food manufacturing industries Australia-wide*

GOODMAN FIELDER • GOOD FOOD GUARANTEED •

means embracing new ideas, working together and setting higher standards. It also means reducing costs, increasing returns and providing a level of customer service second to none.

The nutritional value and safety of food are becoming critical issues for many people. Goodman Fielder has acted on this concern, ensuring its 'Good Food Guaranteed' trust mark appears on all of the company's products, as an indication to customers that the company is committed to producing food of the highest quality – nutritious, good-tasting and safe. The trust mark appears at the top of the consumer panel, listing ingredients and nutritional information, that is printed on the packaging of each Goodman Fielder product.

The visual symbol of this commitment is Goodman Fielder's logo, a stylised quern, or grain-grinding stone. Around for over 4,000 years,

the quern remains the basic food-producing tool in many countries. For the company, however, the quern tells a different story – its contemporary, clean look symbolises Goodman Fielder's heritage as a cereals company, its emphasis on teamwork and unity, its core strengths of innovation and high-quality food production, and its transformation into a world-class food company.

But you don't remain at the top unless you invest in the future. Goodman Fielder keeps ahead of developments in food technology with an international research advisory board composed of food science and technology experts. Substantial investment in research and development (R & D) is also made, with Goodman Fielder spending one per cent of its turnover on R & D each year. Twenty R&D centres are dotted around the globe.

Goodman Fielder's customers have access to many of its R&D centres, such as the Ingredients Centre in Botany, which captures synergies in product development and facilitates problem-solving in applications involving substances such as starch, gelatin and emulsifiers.

The ability to produce products that are not only tasty but nutritious is seen in Nature's Gold™ flour, which has nutrient levels up to 20 times greater than white flour and 10 times more than wholemeal. Another is Hi-maize™, a natural dietary fibre developed by Starch Australasia after 20 years of selective plant breeding. Made from a unique variety of maize, and with the ability to substantially boost the fibre content of food products without altering their taste or texture, Hi-maize™ was launched in 1994 as the essential ingredient in Wonder White, a soft, white bread which is high in dietary fibre.

More recently, Goodman Fielder reached an agreement with McDonald's Australia to include Hi-maize™ in all buns and muffins produced in Australia, significantly increasing their fibre value.

Leiner Davis, along with Starch Australasia and the Germantown operation, is part of Goodman Fielder's global ingredients business. This division integrates 20 manufacturing plants in ten countries, and enables Goodman Fielder to say that if it does not own a brand, there's a good chance that it supplies the ingredients. Apart from leading the global edible gelatin market, the ingredients division is the world's second-largest global producer of specialty ingredients for the dairy and confectionery industries.

Goodman Fielder has placed a high priority on establishing collaborative research programs with external research organisations and is an active partner in several Co-operative Research Centres. These are Australian Government-supported collaborative research ventures bringing together researchers from universities, the public sector and business in the development of innovative technologies.

Overseeing Goodman Fielder's innovation and research functions is the Corporate Technology Group, charged with supporting technology, research, product development, nutrition, food safety, manufacturing and environmental management.

The future interest in the health benefits of so-called 'functional foods' opens a number of opportunities for Goodman Fielder. These are foods that are important nutritionally, and also play a potential role in disease prevention, mitigation and cure. Future food research is likely to focus on the health potential of foods and Goodman Fielder is already responding to these trends with products such as Gold'n Canola, which contains the Omega Benefit, and Uncle Toby's Lite Start.

Goodman Fielder recognises that the challenges facing the food industry into the next century will be the increasing demand for healthy, nutritious and safe food, and for convenience and value.

A wealthy and ageing population is demanding high-quality, value-for-money foods; the number of working mothers is increasing, driving demand for convenient meal solutions; and increasing work pressures are increasing demand for pre-prepared meals and eating out.

Goodman Fielder is focusing on these trends and what they mean for food production in the future, and is in a strong position to exploit them. The company has healthy, staple and high-protein products such as bread, cereals, edible oils and food ingredients and proven innovation skills in developing healthy products such as Hi-maize™, the key ingredient of Wonder White.

Goodman Fielder is establishing a prepared meals business to capture a large slice of the rapidly growing home meals replacement market, which is seen by the company as perhaps the fastest growing and most significant new business opportunity facing the food industry.

Clockwise from above: *Goodman Fielder is committed to putting good food on every table; Starch Australasia is the leading manufacturer of modified starches in Australia; like many of Australia's leading athletes, champion swimmer Kieren Perkins knows the importance of nutritious foods*

Initiatives such as the acquisition of Burns Philp Consumer Foods' sauces business place Goodman Fielder in an excellent position to capitalise on demand for prepared meals.

Goodman Fielder's corporate nutrition policy clearly spells out its dedication to supplying quality food promising a commitment 'to the production and marketing of foods that enhance the well-being and nutritional health of consumers'. The policy says that the company will promote the enjoyment of food and interest in it, relating to a healthy lifestyle.

The Sydney 2000 Olympic Games allow Goodman Fielder to showcase its products – supplied to around 80,000 athletes, officials, volunteers and media, as well as millions of visitors – including cereals, pasta, bread, cooking oils, chicken and snack bars.

Goodman Fielder is also increasingly successful overseas, with exports to the Asia-Pacific region growing. Goodman Fielder International, established in 1996, is responsible for exports from manufacturing operations in the Australian, New Zealand, Papua New Guinean and South Pacific divisions, and Goodman Fielder's Asian investments in countries such as Taiwan. Long shelf-life products, such as noodle mixes for Japan, are being developed specifically for the Asian market.

Australia has an international reputation for producing excellent primary produce. Goodman Fielder is taking those products and turning them into 'Good Food Guaranteed.'

Please refer to page 193 for contact details

State of Victoria

At the south-east corner of the Australian mainland, a fertile landscape of temperate climate has proved itself a rich source of high-quality crops and animal products. Historically, Victoria has relied on the strength of its agricultural and food industries, and in the future they will remain a cornerstone of economic prosperity.

Food accounts for $4 billion per year of Victorian export earnings, but, recognising the still vast untapped potential, the State Government has set an export target of $12 billion by 2010.

Dairy, meat, grains and horticultural products represent the base of the Victorian food industry, but the future is about much more than commodities. It is about converting raw materials into safe, nutritious and diverse food products, processed and packaged to meet the changing demands of world markets.

From less than three per cent of Australia's land area, Victoria's 37,000 farms contribute 23 per cent of the country's agricultural commodities and 30 per cent of its food products. This strong primary industry base has spurred the growth of a large processing sector that now makes up 20 per cent of all Victoria's manufacturing.

The State Government has made the food industry a priority in both its policy decisions and allocation of resources. In the six years since 1992, the food industry attracted more than $1.7 billion in investment. State Government agencies work as a team to create a flourishing environment for businesses, focusing on advances in agri-food science, farm management and practice, food technology, process improvement, value adding and advanced marketing.

Victoria's biggest export markets include Japan, the Philippines, Malaysia, the United States, New Zealand, Taiwan, Hong Kong, Singapore, Thailand and Indonesia. But emerging markets are providing an important addition to the global customer base, with Vietnam, Saudi Arabia, India and Switzerland among many taking advantage of high-quality fresh produce and value-added products.

THE VISION

The Victorian Government envisages a food industry that by 2010 will be recognised as the prime food and food services hub for the Asia-Pacific region and beyond. To achieve this goal, the food sector will become increasingly tailored to customer needs with a focus ranging from niche markets to worldwide food chains.

By 2010 Victoria will reap the benefits of its long-term commitment to research and development, building on a reputation for being innovative, skilled and flexible. Using information technology and stronger communication links, all levels of industry will operate as an integrated food chain.

The aim is to build on Victoria's already renowned food processing sector to create efficiencies second to none. Using state-of-the-art factories and advanced marketing techniques, industry costs will be reduced.

Farmers and rural communities will have thorough knowledge of managing natural assets ensuring high-quality, safe food is grown using environmentally sustainable methods.

The vision is shared by the Government and food industry leaders, with an understanding that by 2010 all links in the industry must be world-competitive.

Clockwise from top left: *Vineyard in the Yarra Valley; Melbourne cafe scene; Victoria Market*

FOOD VICTORIA

Food Victoria was established in 1993 as a framework from where industry, government and the public sector could work together to develop the food industry's potential.

Uniting researchers, growers, processors, packagers, transporters and marketers with government agencies and industry bodies, Food Victoria works towards creating a cohesive approach to the future of the food sector. Its charter includes identifying and removing impediments to growth and market development, creating and nurturing a climate for sustainable growth, and encouraging a world-competitive production and distribution chain, 'paddock to plate'.

The Food Industry Advisory Committee (FIAC), chaired by the Premier, provides Food Victoria with its direction. Also on the committee are the Ministers for Industry, Science and Technology, Agriculture and Resources, and Rural Development, as well as industry leaders and heads of government departments. The Food Industry Taskforce, convened by the Parliamentary Secretary for State Development, ensures prompt government responses to FIAC recommendations.

Top: *Victorian food*
Bottom: *Robins Bush Foods*

KEEPING THE FOCUS ON THE CUSTOMER

Growing high-quality produce and having an efficient food processing sector are only the beginnings of the food value chain. Victorian companies recognise the importance of listening to the needs of a changing market and developing a diverse product range in response to worldwide demands. The aim is simple and it makes good business sense. Satisfy the customer, keep costs as low as possible and be efficient and reliable in the delivery of products. Many Victorian companies are already leading the way. Ardmona, a major grower, processor and exporter of canned fruit and vegetables, with a turnover of $130 million, has developed a quick response to customer needs and opened up niche markets with innovative processed fruit products. It competes successfully in North America, Asia and Europe.

The State Government is helping to foster links between all levels of the food industry and direct connections to overseas markets. A network of eight Victorian Government Business Offices, in the United Kingdom, Japan, Germany, Hong Kong, Indonesia, the United Arab Emirates, the United States and Korea, provide vital links with potential food purchasers, investors and joint venturers.

EXCITING INVESTMENT OPPORTUNITIES

Many of the world's leading companies, including Kraft Foods, HJ Heinz, Nestlé, Cadbury Schweppes, Campbells Soups, Effem Foods and Simplot, have set up regional bases in Victoria. The strong manufacturing sector, creative and well-educated workforce and access to competitively priced raw materials are just some of the features attracting world-class food industry participants. A high-quality business infrastructure, including freight, energy and communications, along with a strong research and development base in agricultural and food sciences and packaging, make Victoria an ideal location for regional headquarters.

Joint-venture opportunities are also providing advantages of lower shipping costs, increased market access, and access to new technology and sources of information. Victoria's Murray Goulburn Co-operative has joined forces with Meiji and Mitsubishi to produce and distribute an internationally recognised infant formula. Meiji supplies the product know-how, Murray Goulburn provides the raw materials and a local manufacturing facility, and Mitsubishi provides worldwide distribution.

Please refer to page 201 for contact details

National Farmers' Federation

Australia's farmers are stacking supermarket shelves from Australia to Asia.

The Asian food market will grow from $750 billion in 1990, to $1,000 billion by the year 2000 – and Australia is in a supreme position to supply a wide range of fresh and processed food products to over 25 Asian countries.

Today, they sell beef to Japan, lamb to Malaysia, oranges to Thailand, and plums to Taiwan. They're also selling yoghurt, ice-cream, specialty cheese and baby formula to the Philippines, Singapore, Thailand and Malaysia.

Tomorrow will see Australia's farmers reaching out even further into Asia, with fresh and processed products, and a lasting commitment to quality control.

The National Farmers' Federation (NFF) will be behind them, every step of the way. As Australia's peak farm body, it represents 120,000 farming enterprises. Its role is to promote policies and issues which will benefit farm families, from trade issues to environmental concerns and the health of country Australians.

The NFF is about to celebrate its twentieth year as a powerful force for the farmers who annually produce around 20 per cent of Australia's export income.

Right: *National Farmers' Federation supports the Australian export industry, with farmers supplying a wide range of fresh food to over 25 Asian countries*

Please refer to page 195 for contact details

Agriculture, Fisheries and Forestry – Australia

Above: *An AQIS meat inspector inspecting a beef carcase*

Opposite: *Loading cattle for export from Darwin*

Australia has built an international reputation as a supplier of safe, high-quality food produced in a clean environment. Whilst responsibility for ensuring the safety and quality of Australian agri-food products ultimately lies with our agri-food industry, government has a vital role in underpinning industry's efforts. Food law in Australia resides with the individual States and Territories that together constitute the Commonwealth of Australia. State/Territory health departments have responsibility for enforcing food standards, while State/Territory agriculture departments are responsible for controlling animal diseases and, in most States, for enforcing the Agriculture and Resource Management Council of Australia and New Zealand meat standards. The Australian Government has a strong legislative base for setting food production and processing standards and for monitoring compliance to ensure food safety. The Australia New Zealand Food Authority (ANZFA), in particular, is a key Commonwealth Government agency with responsibilities in food control.

The Department of Agriculture, Fisheries and Forestry – Australia (AFFA) contributes to this robust, integrated framework which promotes the delivery of safe, high-quality food and sustainable production systems. AFFA seeks to encourage leadership, strategic direction, innovation and support for the Australian agri-food industry with first-class research and well-developed policy and program initiatives, which will take Australia into the next century. The Department performs crucial functions in the areas of quarantine, export and imported food inspection, exotic pest and disease contingency planning, natural resource management, research and development, promoting food safety, and assisting the agri-food industries to adopt quality assurance systems. Some of the specific ways AFFA contributes to enhancing Australia's reputation as a consistent and sustainable producer of food, meeting the highest standards of quality and safety, are described below.

IMPORT RISK ANALYSES

Australia is free of many diseases that could devastate our agri-food industries if they were imported in plant or animal products.

To reduce the risk of this happening, the Australian Quarantine and Inspection Service (AQIS) conducts risk analyses on proposed plant and animal product imports. The analyses are scientifically based and involve identifying and classifying potential quarantine risk organisms and developing procedures to manage them. The analyses meet international standards set by the United Nations' Food and Agriculture Organisation and aim to encourage trade, reduce the use of quarantine as a trade barrier, and protect Australia's plant, animal and public health.

QUARANTINE DETECTOR DOG PROGRAM

As part of its role of administering quarantine, AQIS provides surveillance at airports, seaports, mail exchanges, quarantine stations and other sites, monitoring and examining items that could pose a risk to our agri-food industries.

Since 1991, AQIS has been using 'passenger friendly' beagles for its detector dog program to prevent exotic pests and diseases entering Australia though passengers' luggage or by mail. The beagles are trained to detect different scents, including raw and unprocessed meats, fresh fruits and vegetables, foliage, eggs, reptiles, bees and live birds. When they detect a suspicious odour, the beagles simply sit next to the source and await their reward from the handler.

AQIS continues to expand its detector dog program, and now has teams at all of Australia's major international airports.

In 1995, AQIS introduced 'active' response dogs at international mail centres. These dogs paw and nuzzle suspicious items and are rewarded with a game of tug of war. Three 'active' response teams currently operate in Sydney, at the Mascot Mail Handling Unit, the International Mail Centre, and at private couriers. Similar teams are planned for other capital cities.

EXPORT INSPECTION

Another important AQIS task involves inspecting and/or certifying agri-food exports, such as meat, dairy, fish, dried fruits, processed fruits and vegetables, eggs, grains and organic foods.

AQIS's export inspection arrangements for meat, for example, are recognised internationally as being at the forefront of food safety standards. Australia is the largest international exporter of red meat, exporting to over 110 countries. In recent years, we have consistently had the lowest rejection rate of all countries exporting significant amounts of meat to the United States at its port-of-entry inspection.

All categories of export-meat plants in Australia operate under the Meat Safety Quality Assurance system developed by AQIS. This system is based on International Standards Organisation guidelines and uses the Codex Alimentarius Commission's Hazard Analysis Critical Control Point (HACCP) methodology for process control.

Australia is developing even safer systems for meat production through AQIS's Meat Safety Enhancement Program (MSEP). This involves using company-employed sorters under full-time AQIS

inspection, supervision and veterinary control. It includes detailed monitoring of product hygiene and processes through AQIS's Meat Hygiene Assessment Program.

ENVIRONMENTAL STEWARDSHIP

Sustainable natural resource management is the key to securing the long-term future of agriculture. AFFA is committed to protecting and improving Australia's natural resource base and to achieving sustainable management and investment for future production, processing and delivery by the agri-food industries. Central to this is the integrated research conducted across the Department that contributes to the development of an information base for sustainability, which guides the development of appropriate policies and implementation of effective programs.

In June 1997, the Federal Government established the Natural Heritage Trust with the allocation of $1.25 billion over five years for the repair and replenishment of Australia's natural capital infrastructure. The Trust is a partnership of all Australians, as individuals and as groups, to target environmental problems such as soil erosion, salinity and water pollution at their source. AFFA has a vital role in implementing the Trust through the management of programs, including National Landcare, National Rivercare, Fisheries Action, National Weeds, National Feral Animal Control, and the National Land and Water Resources Audit.

NATIONAL RESIDUE SURVEY

Keeping a check on the level of chemical residues in Australian food commodities is also crucial to our success as a major food-exporting nation and for underpinning our 'clean' image. The National Residue Survey (NRS), located in the Bureau of Resource Sciences of AFFA, monitors, assesses and reports on chemical residue levels in food commodities produced by our agriculture and fisheries industries. The NRS is managed in close co-operation with a range of national, State and industry agencies.

AGRICULTURAL AND VETERINARY CHEMICAL MANAGEMENT

Managing the risks of agricultural and veterinary (agvet) chemicals is a major challenge requiring a robust, integrated and transparent strategy, involving the full commitment of everybody involved in agvet chemical use. This is achieved through the National Strategy for Agricultural and Veterinary Chemicals Management, which further demonstrates Australia's commitment to product integrity and ecologically sustainable development.

AFFA supports best-practice management of agvet chemicals to achieve ecologically sustainable and socially acceptable food production, and supports co-ordinated arrangements between industry and government to better identify chemical uses important to Australia.

SUPERMARKET TO ASIA

The Prime Minister established the Supermarket to Asia Council in 1996 to provide the leadership and drive necessary for export success in Asia. The Council has working groups to address export issues, including the Quality and Safety Working Group (QSWG), chaired by the Minister for Agriculture, Fisheries and Forestry and co-ordinated by AFFA.

The QSWG identified planning for food safety emergencies as a top priority if Australia is to maintain high food safety standards and protect its overseas markets. The QSWG developed a Model Food Safety Emergency Management Plan which has been used by a broad range of commodities industries, including red meat, dairy, horticulture, pork, eggs, poultry, seafood and grains, to develop the sectoral food safety emergency plans.

AFFA also has a wider responsibility in co-ordinating national responses to food safety emergencies. AQIS, in particular, has a vital role in maintaining Australia's access to overseas markets during such emergencies. A key element of this role has been support for the development of enhanced traceability arrangements in those industries most at risk to food safety scares.

The QSWG is also focusing on improving the uptake of Quality Assurance (QA) by the agri-food industries. The confusion surrounding the proliferation of QA schemes and the cost burden to which agri-food enterprises are subjected to meet these quality requirements are impeding the adoption of QA. To address this issue, the QSWG is working with key players throughout the agri-food industry to encourage greater consistency between QA systems and to streamline enterprise-level QA requirements.

QUALITY ASSURANCE FOR SPECIFIC SECTORS

AFFA also works with specific sectors of the agri-food industry to assist them with adopting QA systems. For example, through its involvement with the Grains Research and Development Corporation, the Australian Wheat Board and the Co-operative Research Centre for Quality Wheat Products and Processes, AFFA contributes to research and development strategies to boost the grains industry's international competitiveness. Priorities include developing and adopting QA standards and 'through-chain' product certification. In addition, new wheats and wheat products are being developed to meet the quality requirements of domestic and export markets, and to ensure that the quality of wheat and wheat products remains consistent along the value-adding chain.

LOOKING TO THE FUTURE

These are some examples that demonstrate AFFA's role in enhancing the safety and quality of Australia's agri-food products and in ensuring the sustainable management of our country's natural resource base. The future international competitiveness of our agri-food industry is clearly dependent on Australia maintaining its commitment to environmental stewardship and its reputation as a supplier of safe, high-quality food. AFFA is committed to working in partnership with industry and the community to meet both challenges.

Clockwise from top left: *High-quality Australian fruits; an AQIS detector dog and handler at Sydney Airport; plants undergoing quarantine checks by AQIS scientists at the Plant Quarantine Station in Sydney; a passenger disposing food into a quarantine bin at Sydney Airport*

Please refer to page 184 for contact details

IHD

IHD is a national coalition of independent merchandising companies, combining more than 240 years of experience in rural business. With unequalled access to leading-edge technology and a deep commitment to the long-term interests of the Australian rural sector, IHD offers an unparalleled level of support to all tiers of industry.

IHD was established in 1986 on the premise that all sectors of industry working together should be able to operate profitably, with manufacturers in a position to control the destiny of their own products. The members of IHD, through their own independent business operations, offer professional technical support, research and development (R & D), reseller support and responsible marketing in all areas of crop care, farm management and rural business.

Unprecedented growth in the wine industry has seen wine grape growers taking advantage of the IHD package of services, which include satellite mapping, irrigation scheduling, nutrient monitoring, canopy management and the latest technologies in pest and disease management. Growers, facing increasing international demands, can be confident that they are being recommended the best technology available by IHD.

IHD's team of highly trained professional agronomists is in the field at all times, passing on a first-hand knowledge of Australia's major horticultural production areas and markets. The group's member companies ensure that staff participate in regular training programs to stay at the leading edge of industry developments. Drawing strength from a liberal exchange of ideas and resources, IHD is now recognised as an elite operator, with member companies marketing $900 million of farm inputs, including crop-care products, fertilisers, seed and associated products and services.

Responding to rapid developments in the global food industry, IHD is building on its experience and professional resources. The company believes that many opportunities exist to capitalise on Australia's position as a supplier of high-quality, clean and nutritious food products. IHD is exploring areas of bio-technically enhanced foods that extend shelf life, nutrient enrichment and improvements in cooking and processing.

One area that offers a unique opportunity for Australian producers is the potato industry. Changing lifestyles in Asia and the Middle East have created an expanding demand for many potato

products. IHD is seeking to facilitate this market opportunity by providing improved planting materials, varietal selection and developing a strategic alliance with R & D agencies.

As a major supplier of farm inputs with a broad geographic spread, IHD has a team of dedicated professional agronomists operating in all horticultural production areas. Drawing on a wealth of resources and experience, the company has the capability to operate in a supply-side management role, facilitating the production of quality produce to meet market specifications.

Beyond the potato industry, there are endless possibilities with particular potential in nutraseuticals, where food crops are produced and processed to yield health products that can assist in treating or preventing diseases. In the United States, corn is being grown that will help diabetics control their insulin levels. In Australia and overseas, work is being carried out to increase levels of proteins, specific vitamins and anti-cancer agents in vegetables and fruits. In other developments, wheat is being grown that will produce flour which will extend bread's freshness by up to three weeks.

These are global movements in the food industry that, if harnessed, could represent the future for Australian growers. IHD members have planned study tours investigating the possibilities that are best suited to Australian conditions and potential markets. The group is developing strategic partnerships on a path to exciting advances in food production and marketing.

Top: *A biotechnology researcher at work*

Middle: *Machine-harvesting*

Bottom: *The Australian rural sector is working together through IHD to give unparalleled support to all aspects of industry*

Opposite (top to bottom): *IHD's irrigation technology at work; fresh fruit and vegetable produce; wine grapes benefitting from a range of technological advances made available by IHD*

Please refer to page 194 for contact details

HPLC PUMP

CHAPTER TWO

Technology and Infrastructure

TECHNOLOGY UPTAKE AND AGGRESSIVE INNOVATION

Rapid uptake of new technology and aggressive innovation have been hallmarks of the Australian agricultural community since the first European settlers arrived with their foreign cattle, sheep, grains and vegetable crops.

The European plants and animals were largely ill-suited to Australia's different climatic conditions and soil make-up. The necessity to adapt these plants and animals to Australian conditions has led to a continuing legacy of gene improvement and research and development (R & D), in order to isolate the most vigorous and hardy varieties.

Today, technology plays a vital role in keeping Australia's multi-billion-dollar annual production of agricultural commodities at the international forefront in terms of product innovation, smoother information flow and pro-active export initiatives.

The agri-food industry accounts for around 25 per cent of total manufacturing sector exports. In order to move the products into international markets, the industry is reliant on a complex network of efficient and cost-effective transport systems.

RESEARCH AND DEVELOPMENT

The transformation from traditional agriculture to modern agribusiness has been greatly assisted by the scientific investigations of a number of government and privately funded bodies.

The Rural Industries Research and Development Corporation (RIRDC), for instance, has close ties to primary production and adds around 150 new R & D projects each year to its portfolio of more than 500 projects.

Many of these studies are of direct benefit to the development of agribusiness, and are part of the reason why Australian agricultural research is held in high regard as a valuable export commodity.

As most R & D bodies are looking to commercialise their activities, they tend to work closely with their industry partners.

Above: *Dr Peter Davies, a senior research fellow at the University of Western Australia, using a laptop computer and other equipment to process his field research findings. Research and development initiatives, covering the spectrum from production techniques to land preservation, are helping to transform traditional agriculture into modern Australian agribusiness*

The Commonwealth Scientific and Industrial Research Organisation (CSIRO) is one of the pre-eminent research bodies in the country, contributing to around 600 projects. Many of its programs are conducted with overseas countries, particularly those in the Asia–Pacific region. It also uses its multi-disciplinary skills and facilities to give Australian companies a competitive edge in the global marketplace.

Food Science Australia (FSA) is the country's largest food R & D organisation, and one of the largest in the world. It is pioneering research in areas such as sensory evaluation of manufacturers' products, to compare taste, flavour, colour and size, along with assessing the health potential and inherent design of foods.

FSA has exerted considerable influence in food microbiology and safety, which has given Australian producers an international leading edge in product quality.

INTERNET AND E-COMMERCE

Approximately 20 per cent of Australians are Internet users, although only one-quarter of these users buy and sell over the Internet at present.

As security fears diminish, the number of Internet users increases, and more trading opportunities become available, it is estimated that Internet trade will be as high as 20 per cent of all commercial transactions within the next ten years.

The agri-food industry has been quick to embrace the inherent opportunities involved with this form of commerce.

'E-commerce' on the Internet has advantages at both ends of the spectrum. For buyers, e-commerce streamlines trading and makes ordering a simpler procedure, improving both convenience and savings due to increased competition. Being able to locate competing suppliers of a product and receive useful comparative information will give buyers the upper hand in stretching their budgets further.

For sellers, e-commerce can reduce distribution and marketing costs substantially. It enables marketing to a global marketplace – for many firms, for the first time. These advantages are particularly relevant to businesses, including many farm businesses, operating in rural Australia. It provides opportunities for much closer connection with customers, and for re-engineering business processes to provide more effective customer service.

Above: *Major Australian retailers such as Woolworths are leading the way in modern retail and distribution technology. The advent of e-commerce has major implications for agribusiness. For buyers, it streamlines trading and simplifies ordering whereas, for sellers, it can reduce distribution and marketing costs substantially*

Opposite: *Scientists examining an experimental wheat crop in South Australia. Major research and development organisations such as the CSIRO and the RIRDC tend to work closely with industry, greatly increasing the accessibility of their findings*

NEW TECHNOLOGIES

Satellite communications, the Internet, global news services and overseas travel have brought international commodity markets into focus for every agri-food business.

Australian producers are very conscious that they operate in a global market environment and that, to remain competitive, they have to be at the cutting edge of new technologies, knowledgeable of global market trends and actively developing market opportunities.

The prevalent use of transferable technologies involves taking advantage of the lessons learnt through other industry applications or from simple experience. Many production and processing systems, for any industry, embody a complex array of technologies and knowledge which, for older industries, has been built up over many generations.

The common breeds and varieties used in food and fibre production around the world, for example, wheat and dairy cows, are the result of many centuries of selection for product quality and productivity, pest and disease resistance, and temperament.

DIRECT LINKS BETWEEN RETAILER AND PRODUCER

Growers, processors, retailers and other parties in the agri-food production chain have formed closer relationships in recent times, as the need to develop an export culture has become more apparent.

Schemes such as the Australian Government's Marketing Skills Program, which is designed to improve the international competitiveness of Australian primary industries through the training of networks of producers, have benefited from more than $10 million in funding over the past several years.

EDUCATION

Education is also playing a vital role in increasing the competitiveness of the Australian food industry.

Agricultural education providers such as the University of Melbourne's Faculty of Land and Food Resources – the largest in the Southern Hemisphere – offer courses which have been developed in partnership with industry to ensure increased relevance and best-practice methods.

The University is one of a number of Australian educational institutions which offer direct linkages to market-based programs. Courses offered cover all areas from agribusiness to food science, food processing and food technology.

Plans are also in train to establish national certificates of competency, in an effort to provide employers and workers with improved business-based standards of education and the assurance that skills are transferable.

Clockwise from top: *Australia's extensive network of land, sea and air transport is a vital link in its overall competitiveness, given its proximity to East Asia, a major destination for Australia's exports; Orlando Wines' research laboratory, South Australia; at the University of Technology, Sydney, horticultural researcher Amani Ahmed is taking Australian bush foods to the supermarket shelves after her successful trials of hydroponic growing methods. Australia's agricultural education providers offer courses which have been developed in partnership with industry to ensure increased relevance and best-practice methods. For example, the University of Melbourne's Faculty of Land Resources is the largest institution of its kind in the Southern Hemisphere*

TRANSPORT

Australia's extensive network of efficient land, sea and air transport is a vital link in its overall competitiveness, given its location in relation to East Asia, which is the destination for more than half of its exports.

Considering that many exports into these markets are perishable agri-food items, it would be a waste of valuable promotion and marketing expertise if they did not arrive in the best possible condition.

In 1996, the Prime Minister established the Supermarket to Asia Council to look at ways of improving Australia's agri-food exports to Asia, recognising that Asian countries are the markets for which Australia has an excellent strategic advantage.

Since then, the Council has set up forums, such as the Transport and Logistics Working Group, to develop a comprehensive and strategic action plan. The Group is made up of senior representatives of the transport and logistics industry, from exporters, trucking companies, freight forwarders and logistics companies through to shippers and airlines.

Overall, Australia's rapid uptake of new technology is well suited to the existing infrastructure and will lead the agribusiness community into the next millennium.

Pro-active implementation of currently available technologies, combined with an open mind and a keen eye for future developments, will equip our agri-food industries well for dealing with tomorrow's challenges.

Ben Calvert

State of Victoria

Top: *Airfreight – Melbourne International Airport*

Centre: *Containers being loaded – Port of Melbourne*

Bottom: *Production line – Bonlac Foods Limited*

RESEARCH AND DEVELOPMENT FOR THE FUTURE

The Victorian food industry is aware that its future depends on tailoring its products to market requirements. Whether it be premium tomatoes, trim lamb or customised milk products, all levels of industry from growers to manufacturers are turning to the latest scientific developments to ensure the best product available.

The Victorian dairy industry has found a competitive export edge by developing specific milk-based ingredients with characteristics particularly suited to infant formula, confectionery, biscuits and even pharmaceuticals. Understanding the components of milk has given the industry the capacity to deliver whatever the market wants by way of taste, quality, safety and ease of processing.

The State Government has supported industry in research and development with the establishment of the Australian Food Industry Science Centre (AFISC), designed to solve technical problems and secure market opportunities in a highly competitive and changing market. In a joint venture, AFISC and the Commonwealth Scientific Industrial Research Organisation have formed Food Science Australia (FSA), one of the world's three largest food research and development operations. FSA is a one-stop technology shop for the food industry, with a team of 150 Victorian scientists, engineers and technologists operating from a $40 million complex that includes pilot processing plants.

WORLD-CLASS EDUCATION AND TRAINING

The real strength of the Victorian food industry is the people behind the innovations. With the food sector's heavy reliance on specialist biological, scientific and engineering knowledge, the State Government is committed to linking education to emerging industry requirements. Victoria's world-class universities, Technical and Further Education institutes and the Food Industry Training Board continually develop and update courses for all tiers of industry.

Courses now exist across the spectrum from farm production of plants and animals to produce handling, storage and transport, processed food production, exporting, legal frameworks, catering, sales, management and information technology.

A clever, creative and entrepreneurial workforce with a diversity of skills will continue to ensure that the specific needs of customers are being met around the world.

Please refer to page 201 for contact details

P&O Cold Storage

Food as good as the manufacturer intended it to be. That's what P&O Cold Storage delivers, by providing the longest effective life – from manufacturer to retailer – of food products that display a use-by date.

The storage and distribution of these foods is a science, requiring sophisticated refrigerated warehousing facilities that are strategically located, a modern transport fleet and, above all, state-of-the-art information technology for logistical control.

With over 1.7 million cubic metres of refrigerated space in Australia, New Zealand, the United States and Argentina, P&O Cold Storage is able to ensure the efficient management, storage and timely distribution of food products to major supermarket chains.

In Australia, the company is the largest operator in the distribution and storage of perishable foods business, and in the United States, its presence is growing, with nine warehouses now established nationwide.

P&O Cold Storage activities are divided into three main sectors – produce, manufactured goods and retail distribution – and the company provides services to the frozen foods, chilled foods, manufactured goods, raw materials, dry goods and pharmaceutical markets.

Top to bottom: *P&O's warehouse system is designed for maximum efficiency; conveyor systems are of great assistance in the storage and distribution process; products being loaded for transportation to clients*

Please refer to page 197 for contact details

CHAPTER THREE

A Safe and Superior Product – Pure Quality Assurance

A REVOLUTION IN FOOD SAFETY

Australia's food safety standards have undergone a quiet revolution, the implications of which extend to the kitchens of consumers locally and overseas.

And that is where this new philosophy starts and finishes – the kitchen.

At the end of the day, it is the consumer who drives a food culture so advanced that it has put Australia at the cutting edge of world's-best practice.

Australia's agri-food customers represent a wide range of cultural backgrounds. They recognise the importance of a balanced diet for good health, but also enjoy their food and have sophisticated tastes. They demand quality and variety, freshness and convenience.

Above all, they value safety – and Australian producers have risen to the challenge.

Guaranteeing safe food requires a team effort. The farm, once seen as being isolated from the food industry, is now the crucial starting point of a paddock-to-plate approach to quality assurance.

Australian farmers are pioneering the world's first on-farm quality assurance systems, requiring them to be well informed of the latest developments in animal and plant health, veterinary drugs, chemicals and waste management.

In fact, most farmers like to see themselves as being a part of the food industry.

Gone are the days when livestock was farewelled at the saleyards, or grain was dumped at the depot to be transported to some unknown destination.

Today's primary producers not only know their market, they understand it. Often, their viability depends upon this understanding.

Above: *Australian agri-food is world-renowned, not only for its superior quality, but also for its safety. Ensuring food safety is a team effort – at every step in the food production chain, controls are implemented to eliminate any risk of contamination. Consumer demand for safe food has driven Australia's culture of quality assurance, which is at the cutting edge of world's-best practice*

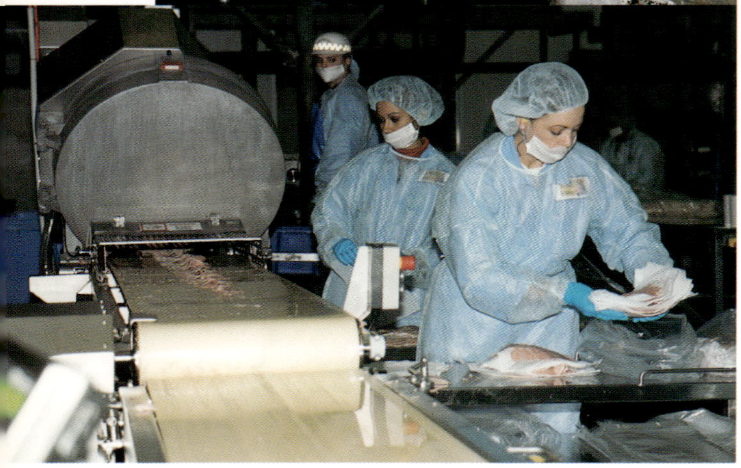

FOOD PROCESSORS AT THE CUTTING EDGE

The market awareness that is evident among primary producers is equally present in the food processing sector. A reputation for safe and high-quality products is a crucial determinant to securing market approval.

The highly trained workforce realises that food safety is a prime responsibility, resting upon everyone who derives a living from the agri-food industry.

While the agri-food industry is ultimately responsible for delivering safe, quality food, the Australian and State Governments support the food companies with an umbrella of agencies that are responsible for regulating uniform sets of standards.

Food law in Australia resides with the individual States and Territories that constitute the Commonwealth of Australia. The Australian Government has a strong legislative base for setting food production and processing standards and for monitoring compliance to ensure food safety.

Several key government agencies oversee the production of food in Australia.

The Australia New Zealand Food Authority sets food standards and labelling requirements, develops codes of practice, co-ordinates surveillance, is responsible for food safety education, co-ordinates food recalls and conducts risk assessment for imported foods.

The Australian Quarantine and Inspection Service (AQIS) is responsible for export inspection, quality assurance promotion, food processing accreditation, import inspection and quarantine.

The Office of Food Safety promotes food safety and serves as a focal point for food safety issues for Agriculture, Fisheries and Forestry – Australia.

HAZARD ANALYSIS AND CRITICAL CONTROL POINTS

At the centre of Australia's new quality assurance culture is a concept known as Hazard Analysis and Critical Control Points (HACCP). The guiding principles of HACCP are promoted internationally as the most effective system of hazard control.

While complex in name, it has a very simple objective – to eliminate the risk at the source.

The HACCP system is based on scientific principles. It identifies specific hazards and the necessary measures for their control, in order to ensure the safety of food. It establishes control systems that focus on prevention rather than reliance on end-product testing.

In order for a quality system to be effective, it must be flexible, and defects have to be prevented at every stage. It is not sufficient to rely on end-point inspection to prevent defects.

Any HACCP system is capable of accommodating change, such as advances in equipment design, processing procedures or technological developments.

HACCP systems actually aid inspection by regulatory authorities and promote international trade by increasing confidence in food safety.

Top and middle: *The appropriate safety garments are worn by all staff handling food at any stage of the food production and distribution chain. Safety caps ensure that the workers' hairs do not fall into the food, and gloves provide a protective barrier against germs and other contaminants*

Bottom: *An aerial view of the feedlot at Rangers Valley Cattle Station. Food processors and primary producers are equally aware of the market's demand for safe, high-quality food*

Opposite: *The Hazard Analysis and Critical Control Points system is at the centre of Australia's new quality assurance culture. It systematically identifies specific hazards and establishes the measures which are necessary for their control*

Grant

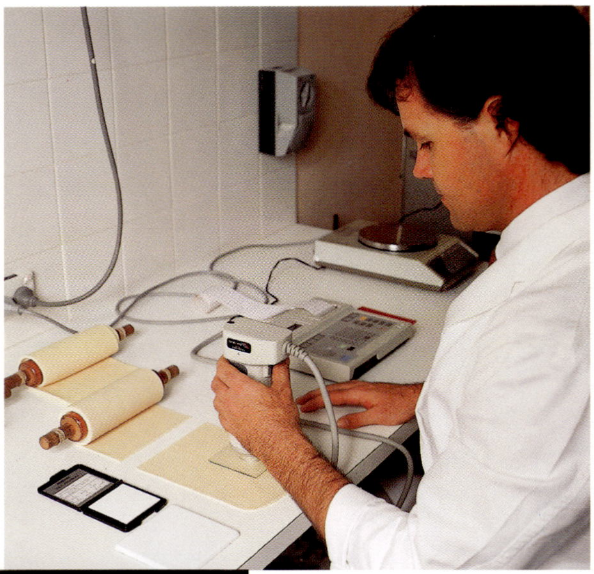

TEAM APPROACH TO QUALITY ASSURANCE

The implementation of a team approach to quality assurance is an inevitable outcome of the adoption of HACCP.

Food safety is no longer the exclusive responsibility of government inspectors. In fact, HACCP-based quality assurance requires food processing companies to convince the regulator that its processing staff have the necessary skills to perform their functions without external supervision.

Modern companies are adopting this team approach to quality assurance. The successful application of HACCP requires the full commitment of management and the workforce.

To achieve this, staff are trained in a spectrum of related areas, including agronomy, veterinary health, production, microbiology, medicine, public health, food technology, environmental health, chemistry and engineering.

HACCP is compatible with the implementation of quality management systems, including the International Standards Organisation (ISO) 9000 series. In fact, it is the system of choice in the management of food safety within such systems.

Many Australian food companies have embraced ISO quality management systems, which provide their products with global recognition of their integrity and, in most cases, a marketing edge.

When seeking export accreditation, Australian food processing companies and manufacturers implement the ISO 9002 quality system or AQIS quality assurance systems that are compatible with ISO requirements. These companies realise that quality assurance is not a one-off commitment.

INDEPENDENT REVIEWS MAINTAIN INTEGRITY

In order to ensure the integrity of quality assurance, Australian agri-food companies are subjected to regular independent reviews. Third-party auditors use standardised schedules to assess whether an HACCP-based quality assurance system is operating in compliance with agreed standards.

Companies and consumers are clearly benefiting from quality assurance. In most cases, there has been a dramatic improvement in the hygiene status of food produced. Companies that are not prepared to comply have often been forced to leave the industry, and consumer confidence is at an all-time high.

The adoption of quality assurance means that food processing companies are determined to supply products to the exact specifications of individual clients, not just to meet the minimum standards established by regulators.

The new arrangements are improving the skills of management and food processing staff, as well as delivering a better industrial environment and better product quality outcomes.

Duncan Brown

Top: *Testing the whiteness of a noodle sheet using the Minolta colour meter*

Middle: *A baker, wearing the characteristic bakers' cap, emptying fresh loaves from bread pans onto a clean surface*

Bottom: *Lab technicians viewing a sample. Australia's food producing and processing companies are subjected to regular independent reviews in order to ensure the safety and quality of all Australian-produced food*

Rangers Valley Cattle Station

Rangers Valley Cattle Station produces high-quality, grain-fed beef exclusively for the discerning Japanese market.

Owned by the Marubeni Corporation of Japan, Rangers Valley is located on a 10,000-acre property in the lush, clean-air environment of the northern tablelands of New South Wales. This area is highly regarded for the production of grain-fed beef due to its high rainfall and fertile soil.

The Angus and Murray Grey cattle used to produce Rangers Valley's superior beef are purpose-bred for the Japanese market by Rangers Valley, or are selected based on genetic superiority from surrounding farmlands. At selection, the cattle are subjected to many checks to ensure that Rangers Valley beef is, and continues to be, of exceptional quality.

Once at Rangers Valley, cattle graze on a feed ration specifically designed by Japanese feed specialists. The corn used in the feed ration is grown at Rangers Valley and is mixed with other quality ingredients under exacting conditions. Rangers Valley uses only the highest-quality natural feed commodities and all feed is tested to ensure that it is free from unwanted chemical residues. The cattle are individually identified and monitored throughout the feeding process, with their performance recorded to assist future cattle selection.

The cattle spend an average of 300 days on Rangers Valley's award-winning feedlot, in a mix of both outdoor and indoor enclosures, where strong environmental regulations and initiatives are enforced. The cattle's gradual weight gain is just part of what is required to produce superior beef for the demanding Japanese market.

The beef is prepared for export to Japan at AUSMEAT-accredited processing plants, with every stage monitored by Rangers Valley and Australian Government inspectors to ensure strict compliance with the customers' needs. The meat is graded into categories according to texture, colour and marbling.

Rangers Valley beef is exported in the form of chilled meat and whole carcasses. Freighted by air, it is in Japanese supermarkets within five days of slaughter, competing successfully in a marketplace that recognises and demands excellence.

Please refer to page 199 for contact details

Top: *Breeding cows on Rangers Valley Farm*

Centre: *All feed is tested to be residue-free*

Bottom: *Rangers Valley high-quality beef*

Food Safety Victoria

Food Safety Victoria, part of the Victorian Department of Human Services, was established in 1997 to ensure that food production within Victoria met national and international safety standards. Food Safety Victoria monitors the status of food from the farm to the consumer, making Victoria's food safety system one of the most reliable in Australia, and providing assurance to domestic and export customers. With exports expected to reach $12 billion by 2010, Food Safety Victoria's role is critical to enhancing consumer confidence in both fresh and processed foods.

Working closely with other government departments and agencies and all sectors of industry, Food Safety Victoria provides leadership across the paddock-to-plate food supply continuum. The agency's aim is to achieve world's-best practices in managing food risks through a co-operative, co-regulatory framework.

Food Safety Victoria monitors compliance to food legislation and provides a rapid and co-ordinated response to public health emergencies. The agency contributes to the development of food standards, food industry guidelines and codes, and focuses on establishing and maintaining international recognition for Victoria's food safety alliance.

Top: *A modern Melbourne bakery*

Above: *A Melbourne biscuit baker at work*

Below (left to right): *Consultancy services being provided by Food Science Australia to food processors; pilot scale processing at Food Science Australia; a food inspector checking the temperature in a commercial refrigerator*

Please refer to page 201 for contact details

Darling Downs
Bacon Co-operative

Natural pork bred and processed under tightly controlled quality conditions is the essence of Darling Downs Bacon (DDB) Co-operative.

The Co-operative's suppliers and processing facilities are governed by a strict quality assurance regime, fully government approved and regularly audited. This ensures the highest safety standards in smallgoods production and presentation.

'Everything we do is targeted to meeting customer needs', says a DDB spokesperson.

'We are a totally Australian-owned company committed to our farmers and staff, and also committed to best practice in our industry.

'Our high standards, coupled with our market-driven focus, allow us to supply the changing needs of our clients with the best product available.'

DDB's single revitalised processing plant, based at Toowoomba in the rapidly growing south-east corner of Queensland, and its well-established network of agents and distributors, mean that DDB is well placed to dominate its chosen market.

Improved production techniques at the Toowoomba plant have given the Co-operative the confidence it needs to take on the new century with customer-oriented lines and top-quality service.

A vertically integrated organisational structure has allowed the Co-operative, close to 90 years old, to retain its traditional rural strength. With 169 working pig producers as its exclusive shareholders – around half of the total pig producers in the State of Queensland – DDB can move aggressively into the 21st century with a market-focused, export-oriented approach to its products and customers.

These products include fresh pork and processed ham, bacon and smallgoods.

Major hospitals, universities and airlines across the nation trust and rely on DDB quality assurance, taste and appeal, as do corner-store retailers and even bakeries. This trust, along with an annual turnover of $A119 million, has resulted in DDB being ranked among Australia's Top 400 businesses.

Please refer to page 190 for contact details

Top: *The fertile Darling Downs region*

Bottom: *A delicious platter of ham*

CHAPTER FOUR

Sustainable Solutions

SEEKING BALANCE IN A LAND OF CONTRASTS

Australia is a land of powerful environmental contrasts which add to its allure, but must not be taken lightly.

On one hand, it is a land of plenty. Having a relatively benign and temperate climate, with seemingly unbounded natural resources, Australia is one of the few countries in the world that can be readily self-sustaining.

On the other hand, it is a fragile ecosystem which is marked by extremes. Australia is the driest continent outside Antarctica, and much of its area is covered by desert. However, not far over the mountains from the harsh plains, the steamy coastal hinterlands are drenched with up to 7,500 millimetres of rain per year.

Nature's delicate balance is taken into consideration by modern farming techniques, which attempt to preserve natural resources, instead of simply rewarding the quest for higher and better production.

Governments, industry bodies and individuals are contributing to the conservation effort by implementing pro-active policies to address any imbalances and to ensure that Australia's positive image and levels of output are guaranteed for the future.

With the increased level of urbanisation and industrialisation, the appropriate disposal of effluent is becoming a top priority. Where it was once common for industry to release effluent straight into the environment, regulatory bodies now enforce some of the most stringent environmental laws in the world.

The high standard of water, soils and environment is not only an aesthetic imperative – it is also a major contributing factor to the appeal of Australia's agri-food commodities in international markets.

Above: *Revegetation projects such as this one are taking place all around Australia, as governments, industries and individuals play their part in a combined effort to maintain the country's delicate ecological balance. The purity of Australian soils, water and environment is not only an aesthetic imperative, but also a major reason for the universal appeal of its agricultural produce*

FARMERS AND RESEARCHERS WORKING TOGETHER

Commonwealth Scientific and Industrial Research Organisation (CSIRO) land and water researchers are at the forefront of scientific efforts to maintain the environment, repair existing damage and devise strategies to prevent future harm from occurring. Their work is being supported at ground level by food producers, who are using the valuable research information in league with their own initiatives.

The CSIRO is one of the largest and scientifically best qualified environmental research agencies in the world. It tackles issues which are central to the future prosperity of Australia, its communities and industries – such as salinity, acid soils, soil erosion, loss of land fertility, pollution and water quality.

CSIRO Land and Water is also developing its environmental export sector, providing technology, knowledge and advice to overseas customers who are struggling to cope with the problems of pollution and degradation caused by rapid development.

At a local level, groups like Landcare are taking a community-based approach to fixing environmental problems and protecting the future of our natural resources. Australia has more than four thousand Landcare groups, whose members include one in three farmers.

Landcare aims to rectify or prevent problems such as the spread of weeds, increasing soil salinity and the expanding numbers of feral animals. It uses whole-farm planning methods within individual enterprises, along with programs such as Waterwatch, Watertable Watch, Streamwatch, Salt Watch and Worm Watch, on a broader scale.

GROUNDWATER IN FOCUS

Groundwater is currently a major conservation topic, with the Federal and State Governments having initiated sophisticated maintenance and development policies for catchment authorities.

Above: *Daybreak at Koolmoon Creek in far north Queensland. Australia's waterways are not only breathtakingly beautiful but also precious natural resources, often forming the basins for major agricultural regions. For example, the Murray–Darling Basin itself is about the size of France or Spain*
Opposite (clockwise from top left): *Research into plant water use and groundwater monitoring gives cutting-edge information on recharge control and the effectiveness of various vegetation types in controlling salinity – some plants survive better than others in high-salinity regions; discussing treatment solutions for soil erosion – revegetation requires a whole plan of management, encompassing stock rotation, fencing and weed removal; measuring aspects of stream health in a cleared pasture in the Mary River Catchment, south-east Queensland*

Groundwater issues have the potential to affect a large number of productive Australians, particularly outside the cities, in the major food-producing regions.

Authorities have consulted widely with farming and agribusiness communities, which are the largest users of groundwater and also the hardest hit by dry seasons, to develop effective strategies.

Much of the responsibility for groundwater management has been put back into the hands of local users, who take a direct interest in the protection of this resource.

CORRECT MIX OF ENTITLEMENTS

Getting the mix of irrigation entitlements right is crucial in ensuring that aquifers, the porous rock layers which retain vital subterranean supplies of water, are able to be replenished.

This delicate balance is essential to the long-term viability of agriculture in Australia. The replenishment of aquifers is in the vicinity of several million megalitres

Department of Agriculture
and Rural Affairs

per year across the country, but it is important to remain vigilant regarding the amount of water which is being extracted. Hence, the allocation of licences will be largely determined by the need to maintain this balance.

Allocations are particularly relevant, as water is the lifeblood of around 7,000 farmers and 40 towns, many with industries relying on water, which line Australia's largest river system, the Murray–Darling.

To give an indication of the size of the area which feels the impact of this system, the Murray–Darling Basin itself is about the size of France or Spain.

Inadequate groundwater management can lead to rising salinity and algal bloom, along with non-economic factors such as a reduction in native animal and plant life, contraction of forests and degradation of other wetland habitats.

PRESERVING OUR ENVIRONMENTALLY RESPONSIBLE IMAGE

New and emerging industries, such as aquaculture, benefit greatly from Australia's environmentally responsible image on the international stage.

Uncontaminated water in a pristine environment is a significant marketing factor, which is not available to enterprises in other parts of the world.

Many overseas areas have been damaged by the effects of land and water degradation, industrialisation and population pressures, which have a trickle-down effect through their domestic agri-food industries.

The top priorities include managing soil and water resources for sustainable production and, at the same time, keeping clay nutrients and organic matter out of water catchments, preventing soil acidification, improving environmental quality and using practical tools to monitor progress.

SOLVING THE SALINITY ISSUE

Salinity was one of the earliest and most visible effects which prompted the development of better management practices.

One research initiative on the Murray River proposes that millions of tonnes of salt be stored in specially designed ponds in irrigation areas, which would protect rivers and freshwater lakes from the threat of salinity.

Huge quantities of salt would be involved, with an estimated one million tonnes being diverted from reaching the Murray River annually.

Clockwise from top: *This wetland habitat survey project, aimed at protecting the habitat and researching its vegetation to assist the survival of local native animals, was the winner of Western Australia's Catchment Award; the Katanning Creek Catchment Group won the Catchment Award in the seedling category by collecting seeds and propagating local native plants for community landcare projects; Landcare groups use satellite maps to find areas where salt is rising – the patchy ground shows where the water table is rising, an indication of salinity*

MANAGING NUTRIENTS

Nutrients are helpful agents in promoting plant growth and are important in the natural food chain. However, in large amounts, they can cause excessive algal growth, including blue-green algae, in waterways – which can put natural ecosystems out of balance, harming water-life and animals.

Algal growth can also affect human uses of water for purposes such as drinking, recreation, stock water and irrigation.

Various schemes are providing landholders with opportunities and information to deal with land degradation problems and to improve land management practices.

Initiatives taken to date include employing conservation farming techniques, actively controlling soil erosion problems, protecting local wetlands for their role in filtering sediment and nutrients, and providing a diversity of wildlife habitats. In addition, constructing farm dams encourages water-plant growth, so that they act as filters for run-off. Another effective conservation strategy is to re-use effluent safely.

TRADABLE WATER ON THE AGENDA

Tradable water is one of the hottest issues in the rural sector.

Traditionally, land and water have been inseparable, with water rights attached to the property to which they relate. This has changed recently, so water authorities are looking to increase the transparency of the trading system, which will be a boost to the agri-food industry. It will mean an increase in efficiency, as water is diverted into the most flexible and profitable areas.

Highly profitable industries will be able to purchase part of the available water assets, which may otherwise have been directed to less lucrative enterprises.

It will also benefit the rural sector by opening the way for the sale of water to industries in rural or urban centres and to farmers for stock drinking supplies.

All this is part of giving Australian agribusiness a firm platform from which it can produce some of the world's best food, well into the future.

Ben Calvert

State of Victoria

Above: *Victorian food*

A COMMITMENT TO SUSTAINABLE DEVELOPMENT

The Victorian food industry recognises that its future relies on clean water, fertile soil and a balanced ecosystem. A range of joint initiatives between the State Government, farmers and businesses have been established to ensure a balance between a healthy environment, using land for agriculture, and providing water for drinking, industry, irrigation and other agricultural uses.

The State Government has established the Sustainable Agriculture Program to help farmers gain the skills and information they need to manage natural assets and to use natural resources efficiently with minimal environmental impact. Some of the projects that have provided practical support include minimising the run-off from irrigated pastures, sustainable management of dairy shed effluent and the development of integrated pest management strategies to minimise chemical pest control in horticultural industries.

The State Government has also recognised the economic and environmental benefits of reducing waste from food processing industries. A strategy to develop cleaner production initiatives also promotes reuse and recycling. Part of the commitment is to provide interest-free loans for installing proven clean technology or waste audits, and grants for process and system improvement to minimise waste.

In the area of weed and pest management, the State Government has set a target to reduce the incidence of weeds and pests by 90 per cent on treated land, and minimise the impact on economic losses and the integrity of native ecosystems. An allocation of $2.5 million a year has been made to the Good Neighbour program, which encourages landowners and landcare groups to control a variety of weeds and pest animals.

Other State Government initiatives include:

- a groundwater management strategy;
- tradable water, to facilitate the best use of water resources;
- a nutrient management strategy; and
- salinity management plans,

where farmers are adopting better irrigation practices, pumping groundwater, planting trees and deep-rooted perennial pasture species, as well as constructing drains and implementing whole-farm plans.

These measures, when combined, will secure the sustainable production of food in Victoria for generations to come.

Please refer to page 201 for contact details

Business Ballarat

When the Japanese noodle manufacturer, Hakubaku, began scouting for a location to produce organically grown wheat noodles for export, the search ended in Ballarat, Victoria. Close to the Port of Melbourne and Melbourne Airport, Ballarat has excellent distribution channels, a stable workforce and a clean environment for the production of accredited organic foods.

The Hakubaku investment is putting Ballarat noodles on the tables of Japanese diners, and has joined premium cool-climate wines, patés and muesli bars as part of the rapid growth in gourmet and niche market food and beverages produced in Ballarat. The city has proved itself to be an ideal processing and export base for an elite group of multinational food producers which have long-established operations, including Mars Confectionery, McCain Foods and Bunge.

Business Ballarat, the economic development arm of the City of Ballarat, is assisting companies to make the move to Ballarat and promoting the city's many advantages as a location for food processing and value adding to regional produce.

Please refer to page 188 for contact details

A selection of gourmet delights produced in the Ballarat Region

CHAPTER FIVE

A World-Class Supplier

SUPERIOR PRODUCTS AND KNOWLEDGE

A growing reputation for tailor-made quality and consistency is spearheading new avenues of world-class development within the Australian food industry.

Whether it is exporting prime lamb to the restaurants of New York, lobster to the dinner tables of Japan or wine to the bottle shops of Europe, Australia's reputation as a world-class, large-scale supplier of fine fare is growing.

Building upon an enviable international record of being one of the world's cleanest food production environments, Australia is increasing the value of its exports through the application of technology, knowledge and skill.

The term 'clean and green' has long been accepted into the lexicon of food production – so much so that today, it is taken for granted that anything that emerges from the Australian food production cycle is produced in a sustainable and ecologically friendly manner.

A commitment to bettering international benchmarks and fully integrating supply chains is helping the industry to compete with other countries, which may be concentrating on price considerations alone.

INNOVATION, QUALITY AND CONSISTENCY

For Australian agribusiness to compete in the international marketplace, it must secure growth through innovation, product quality and consistency.

A crucial platform – and logical extension – of the nation's push to secure higher value markets for Australian produce is the Federal Government's Supermarket to Asia (STA) program.

Established by Prime Minister John Howard, the initiative aims – through improved dialogue – to capitalise on Australia's proximity to emerging Asian markets by identifying areas of strategic advantage. It has worked to overcome both political and physical barriers in order to increase the level of Australian exports to the region.

Above: The Supermarket to Asia program aims to capitalise on Australia's proximity to emerging Asian markets by identifying areas of strategic advantage and improving dialogue between the various parties involved. It acts as an information conduit for Australian companies wanting to find out more about trading within the Asian region, as well as providing a forum for those companies already working with Asian customers to benchmark their performance

STA is now a conduit for Australian companies wanting to find out more about trading within the Asian region. It also provides a forum for those companies already working with Asian customers to benchmark their performance through the flow of information and ideas.

While STA remains an important link in keeping the industry aware of Asian market developments, a silent information-based revolution is also taking place on farms and in factories and trading houses around the nation.

COMMUNICATIONS OVERCOME DISTANCE

In the past, distance was the traditional barrier to the flow of information across a continent as vast as Australia. Today, it is no longer a problem, with telecommunications providers linking the trading houses of the world to companies producing and manufacturing products across the nation.

Internet and e-mail access to all manner of information, from trends to daily market reports, is now complementing information delivered by facsimile machines and postal services.

The range of products available is also more diverse than at any other time in the country's history, with niche services and publications being tailored to specific aspects of particular industries and also to emerging sectors.

The Kondinin Group, a non-profit Western Australian organisation, is representative of a movement to supply Australian agribusinesses with credible, results-focused data. The Group's *Farming Ahead* publication combines up-to-date research and development findings with product test reports and information on international markets.

Above: *Sprinklers like this are keeping Australia's fertile winegrowing regions well watered*

Opposite: *A vineyard at Giants Creek in the excellent Hunter Valley winegrowing region. The relatively recent international success of the wine industry illustrates the ever-expanding range of world-class agri-food products for which Australia is renowned*

Overleaf: *A school of trevally swimming in the clean, clear waters off eastern Australia. Australia's seafood industry accounts for around five per cent of its rural exports, with the total value of production exceeding $1.7 billion. The industry provides direct employment for about 20,000 people*

TARGETING INFORMATION FLOW

Increasingly, commodity organisations are using targeted publications, both electronic and hard-copy, to develop their own identities and databases. Such services are seen as being complementary to existing media services provided, largely, on a national weekly basis by the Rural Press organisation.

Other communication gaps are being filled by commodity-specific publications, such as *Wool Day* and *Beef Day*, which are available via facsimile or e-mail.

The use of consultants and other third-party agencies is helping to focus industry attention on the need to look at meaningful change in both the provision and usage of information, as well as the adoption of new management and production practices.

Education is also playing a vital role in increasing the competitiveness of the Australian food industry.

Providers such as the University of Melbourne's Faculty of Land and Food Resources – the largest in the Southern Hemisphere – offer courses which have been

KING ISLAND DAIRY

developed in partnership with industry in order to ensure increased relevance and best-practice methods.

Australia's research and development industry is perhaps the most important in linking the agri-food sector with consumers and in ensuring a sustainable future.

Covering everything from land degradation to sensory perception, the national research effort is widely regarded as being world-class. As a result, many Australian inventions and ideas are now being marketed internationally.

The development of bodies such as Food Science Australia – a joint venture between the Commonwealth Scientific and Industrial Research Organisation's Division of Food Science and Technology and the Australian Food Industry Science Centre – is helping manufacturers and exporters 'get it right' from the beginning.

The Rural Industries Research and Development Corporation, an equally important research and development body, but with closer ties to primary production, is also a world-class provider of outcomes which are crucial to Australia's standing as a front-line exporter.

SEAFOOD IS A BENCHMARK

The seafood industry typifies the overall agri-food industry's aim to develop and cultivate new markets.

With seafood being one of the most internationally traded commodities (20 million tonnes per year) and Asia being the dominant player in the market, it is a sign of the Australian industry's dedication that it continues to play a significant role in the Asian marketplace.

Seafood is a commodity where product quality and consistency are paramount, and Australian research and development bodies, producers, manufacturers and exporters have worked diligently to satisfy and exceed the market's expectations.

Australia now supplies about $1.1 billion worth of seafood per year to most Asian countries. The Asia–Pacific region accounts for about 90 per cent of the value of all seafood exported.

Such market gains have come through persistent and strategic alliances. These alliances, no matter what the commodity, have been fostered by umbrella marketing organisations such as the Australian Prawn Promotion Association (APPA).

Typical of similar organisations working internationally to raise awareness levels, the APPA is active in the industry's main target markets, staging exhibitions and fostering promotional alliances with supermarket chains, such as the Japanese Jusco group.

Australia's geographic location within Asia and reputation as a world-class supplier is proving attractive to international agri-food businesses looking for a base within the region.

Top and above: *A wealth of up-to-date information is available through publications such as the* Australian Farm Journal *and* Town and Country Farmer. *These and other publications, often available in electronic and hard-copy formats, form part of the vital information links that are keeping Australia at the forefront of the international agri-food industry*

Opposite (clockwise from top left): *Education providers such as the University of Melbourne's Faculty of Land and Food Resources – the largest such faculty in the Southern Hemisphere – offer courses that were developed in partnership with the agri-food industry, ensuring relevance and best-practice teaching methods; a selection of premium Australian food; the Ord River Irrigation Scheme, Packsaddle Plain, Western Australia. Western Australia's Kondinin Group, a non-profit organisation, represents a growing movement to supply Australian agribusinesses with credible, results-focused data*

Above: A Thai prawn salad prepared by chefs at the Hotel Inter•Continental, Sydney, using fresh Australian ingredients. The ready availability of top-quality produce, excellent growing conditions, a well-educated labour force and proximity to Asia are among the many reasons why agri-food multinationals are choosing to invest in Australia

MULTINATIONALS BASED IN AUSTRALIA

In the last decade, numerous multinational corporations have established regional headquarters in Australia, with a view to adopting a strategic position within the region.

Multinational DanPork is building the nation's largest specialist abattoir, worth about $57 million, at Pratten in Queensland's Darling Downs region. It will process approximately 800,000 pigs per year for the Asian market. The company expects to send at least 80 per cent of its production to Asia, focusing on the premium markets for chilled pork. Long-term contracts will be offered to farmers to give them the confidence to expand and to supply the additional pigs that the company will need to meet this aim.

The Italian-based Parmalat Group, employing some 25,000 people in 21 countries, and with an annual turnover of approximately $8 billion, chose to establish its Asia–Pacific regional headquarters for manufacturing and research and development in Australia. Parmalat is one of the world's largest processors and distributors of food, including dairy products, fruit juices and nectars, pasta, sauces, desserts, chocolate, pastries, biscuits and bread-based products. Parmalat currently owns and operates dairy-food manufacturing facilities in Albury and Wodonga. Over the next five years, the company intends to invest several hundred million dollars and create around 300 new jobs.

United States-based food conglomerate Cargill, with 44 lines of business in 72 countries, is another company which is committed to developing Australian agribusiness. Cargill's operations are typical of the strategic positions now being taken by multinational food corporations, as the industry becomes more competitive than ever.

From humble beginnings in the late 1960s as a one-person trading operation, Cargill Australia has pursued a path of strategic expansion. Today, it is a key player in beef processing, oilseed crushing, grain trading, salt production, fertiliser distribution and juice merchandising. The firm has played an important role in helping the Australian beef industry to implement best-practice food safety and processing efficiency standards, paving the way for greater international market access for the nation's cattle producers.

Frequent assessment of Australia's potential continues to show that of all the countries capable of supplying food to the growing Asian market, Australia continues to stand out with its abundant natural resources, large land base and skilled workforce.

If, as all indications suggest, Australian agribusiness is to continue nurturing and adding economic substance to its image as a world-class supplier, it will be through the co-operation of all sectors within the industry to deliver products of the highest quality and consistency.

David Gorrie

Maxi-Cube

When a banana picked in northern Queensland is destined for fresh fruit markets in the south, it could, on any one journey, encounter every type of weather from tropical heatwaves to snowstorms. With consumers now demanding produce as near to natural as possible, the transport industry has come to rely on Maxi-Cube, Australia's leading brand in temperature-regulated transport equipment. Whether it be strawberries from Victoria or tuna from the New South Wales coastline, Maxi-Cube's refrigerated trailers ensure that Australia's often harsh and varied climate does not spoil the quality of our fresh produce.

Founded in 1972, Maxi-Cube, now known as MaxiTRANS Industries Limited, is the largest manufacturer of trailers for Australia's transport industry. Providing equipment to suit individual requirements, Maxi-Cube trailers maintain specified temperatures from -29 degrees Celsius to two degrees Celsius. Committed to continuous improvement and development, the Maxi-Cube Design Centre uses Computer-Aided Design Systems to reduce design costs and increase delivery speed to the market. Maxi-Cube's most recent innovation is the dual-temperature trailer, which is capable of transporting frozen and chilled freight simultaneously. One trailer can do the job of two or more, with retail outlets and consumers benefiting from cheaper transport costs and delivery convenience.

The purchase of Freighter Australia in 1998 expanded the company's product range well beyond refrigerated trailers, with Freighter being renowned for its curtain-sider trailers. The combined company now has annual sales of $100 million and is able to meet the requirements of a much broader market base. Leading transporters, including Lindsay Bros, Refrigerated Roadways and Linfox, carriers for Woolworths and Coles Supermarkets respectively, entrust their fresh freight to Maxi-Cube, as do hundreds of owner-operators around the country.

Maxi-Cube's commitment is to listen carefully to the needs of customers and live up to delivery promises. From the first inquiry to technical support, design evaluation, delivery and operations testing, Maxi-Cube guarantees a comprehensive customer service and contributes significantly to the enjoyment of Australia's natural produce.

Above: *Maxi-Cube Dual Temperature Refrigerated Trailer*

Below: *Maxi-Cube Refrigerated B-Double Combination Trailer*

Please refer to page 194 for contact details

Cargill Australia

Top and above: *Cargill has brought its world-class standards of food safety and efficiency to its beef processing business, opening the way for greater export opportunities for Australia's cattle producers*

Opposite: *From its beginnings in Australia's young oilseed crushing industry in 1972, Cargill has grown into the major oilseed crusher that it is today. Cargill brought its considerable expertise into establishing canola as a major Australian crop*

Since Cargill began business in 1865 as a grain warehouse in Iowa in the United States, it has grown into one of the world's largest privately held companies, merchandising, processing and distributing agricultural and other essential commodities and services across the globe. Cargill now has 44 lines of business in 72 countries and employs some 79,000 people in over 1,000 locations.

Success has been built on the basic beliefs of ethical business practices, long-term vision and world-class standards. Quality products, services and relationships of trust with customers and suppliers are essential components of this philosophy. No matter where it does business, Cargill maintains the same high standards for technology, worker safety, food safety and environmental safeguards.

Cargill's vision is to be the best at merchandising, processing and distributing agricultural and other commodities. It has the strengths of a multinational company, but Cargill Australia is managed and staffed by Australians. Cash-flow generated in Australia is reinvested to provide products and services for customers and offer rewarding careers for employees.

From a one-person trading operation in 1967, Cargill Australia has grown to include businesses in salt production, oilseed crushing, beef packing, fertiliser distribution, grain trading and juice merchandising. Nevertheless, it operates as one company, gaining unity and efficiency by sharing common support systems in information technology, accounting, finance, purchasing, maintenance and human resources.

At the farm gate, Cargill presents one face to farmer-suppliers. Field officers work for both the Commodity Marketing and Oilseed Divisions. As well as providing agronomic advice and quality tests, arranging storage and scheduling transportation for farmers, they also source oilseeds for Cargill's crushing mills and grain for Cargill's beef feedlot.

From its beginnings in Australia's young oilseed crushing industry in 1972, Cargill has grown into the major oilseed crusher in the country, and a recently built plant in Newcastle, New South Wales, enhances Australia's opportunities in the vegetable oil industry. Working closely with farmers and cotton ginners, Cargill's commodity marketing and oilseed crushing businesses have played a pivotal role in the success of Australia's cotton industry. More recently, Cargill brought its expertise to establishing canola as a major Australian crop.

In Western Australia, Cargill operates the second-largest evaporative salt field in the country, supplying salt of extremely high purity to industrial customers throughout Asia.

In beef processing, Cargill brought its world-class standards of food safety and efficiency to the industry, opening the way for greater export opportunities for Australia's cattle producers. Through the transfer of worldwide expertise, Cargill delivers customer focus and adds value to producers' operations through supplier partnerships.

Cargill has thrived for more than 130 years by managing risks effectively, especially the risks of price changes in commodity and financial markets, and by thinking globally, while meeting local customer needs. It has built expertise in sourcing, storing, trading and managing the logistics of large volumes of commodities and products, building and operating facilities at low cost to process raw materials into value-added products of high quality. Cargill has built a solid, profitable group of businesses in Australia and sees tremendous long-term growth ahead for the Asia–Pacific region, in which Australia will play a key role.

Of all the countries capable of supplying food into Asia, Australia has the greatest advantages – skilled labour, a huge land base and abundant natural resources. Cargill will work closely with customers and suppliers as Australia uses its competitive advantages to develop export markets. As the world's consumers increase their standards for quality and food safety, Cargill and its customers will be well placed to serve the new demand.

Please refer to page 188 for contact details

Ridley Corporation

Top: *Ridley AgriProducts Terrang Mill, Victoria*

Centre: *Cheetham Salt Limited*

Below: *Ridley Canada Limited's 'Cotswold Pig'*

Established in Australia in 1987 and now operating internationally, Ridley Corporation is Australia's largest stockfeed miller and salt producer/refiner, and through its 60 per cent-owned subsidiary, Ridley Canada Limited, among the top 10 producers of stockfeed in North America.

Its Australian stockfeed business, Ridley AgriProducts, comprises a network of 23 stockfeed mills across Australia, producing and distributing a comprehensive range of animal feed products for the beef, dairy, pig and poultry industries, as well as for horses and sheep. The Pet Products division manufactures and markets specialty aquafeed, birdseed and dog food products.

In North America, Ridley Canada is one of three stockfeed producers with significant operations in both Canada and the United States, operating 35 stockfeed production facilities in major livestock production regions. Through its subsidiary, Cotswold Canada, Ridley Canada operates a swine-breeding business and has exclusive rights to market the genetically advanced line of Cotswold pig. All the Canadian operations are accredited to demanding ISO standards.

Cheetham Salt has a major share of the Australian refined salt market and also produces salt for use in processed foods, stockfeed, pet food and industrial applications. Cheetham distributes salt products throughout Australia and the Asia-Pacific and has a 50 per cent interest in Dominion Salt, New Zealand's largest salt producer.

One common theme unites all four Ridley businesses – a commitment to quality, customer service and product development. The stockfeed, supplement and premix products manufactured are the results of extensive research and development and are formulated to provide consistently high-quality nutritional formulations that enhance animal health.

Ridley's experience in achieving optimum nutrition with its feed products can be tapped into by stock producers. Ridley Agriproducts provides an animal nutrition assessment service, whereby the company will individually examine an animal's conditions, such as living environment, age and food supply, to tailor a correctly balanced animal feed.

Ridley Agribusiness' own proprietary brands have been developed with the same care, and cover a variety of animal feeds.

Please refer to page 200 for contact details

Wirrilla Pty Ltd

When a food supplier is a highly successful exporter to the exacting Japanese market, you know that its produce must be of the highest quality. Wirrilla Pty Ltd, growers of fresh asparagus, have been supplying this market for several years, seeing sales to Japan soar from 300 tonnes in 1989 to over 700 tonnes in 1997.

Established in 1956 and located on a 1,200-hectare property at Jugiong, near Yass in southern New South Wales, Wirrilla is a family-owned company involved in all stages of the growing, packing and distribution of fresh asparagus. Its main market is Japan, but it also exports to Singapore and Hong Kong and supplies the domestic Australian market, with Woolworths being the key customer. In 1996 the company was named New South Wales Exporter of the Year for Agricultural Products.

The recent establishment of a second property at Katherine in the Northern Territory has expanded Wirrilla's capacity and enabled it to lengthen the period in which asparagus is available. David Power, Wirrilla's Managing Director, is aiming to extend current seasonal supply so that the company can supply asparagus in periods when world levels of asparagus are low.

Wirrilla has also broadened its crop base, growing mangoes, lemons, red-fleshed grapefruit and annual fruits and vegetables at the Katherine enterprise, and establishing a vineyard at Jugiong that will eventually supply both the domestic and export markets. Asparagus and wine production at Jugiong are being integrated to maximise efficiency and economies of scale.

Behind Wirrilla's success is a business ethos based on four tenets: quality, communication, honesty and relationships based on friendship.

Sustainable quality is seen as paramount and pervades all stages of the company's operations from planting to supermarket delivery. As part of Wirrilla's quality assurance, produce is monitored during its road and air transport, ensuring that optimal temperatures are maintained.

As Wirrilla enters a growth phase, it will maintain its competitive advantage through its successful mix of product quality, reliability, adaptation to customer requirements and superior innovation.

Please refer to page 204 for contact details

Top: *Fresh Wirrilla asparagus, destined for the export and domestic markets*
Above: *The beautiful Jugiong region of southern New South Wales provides the setting for both asparagus and wine production*

Map of Australia

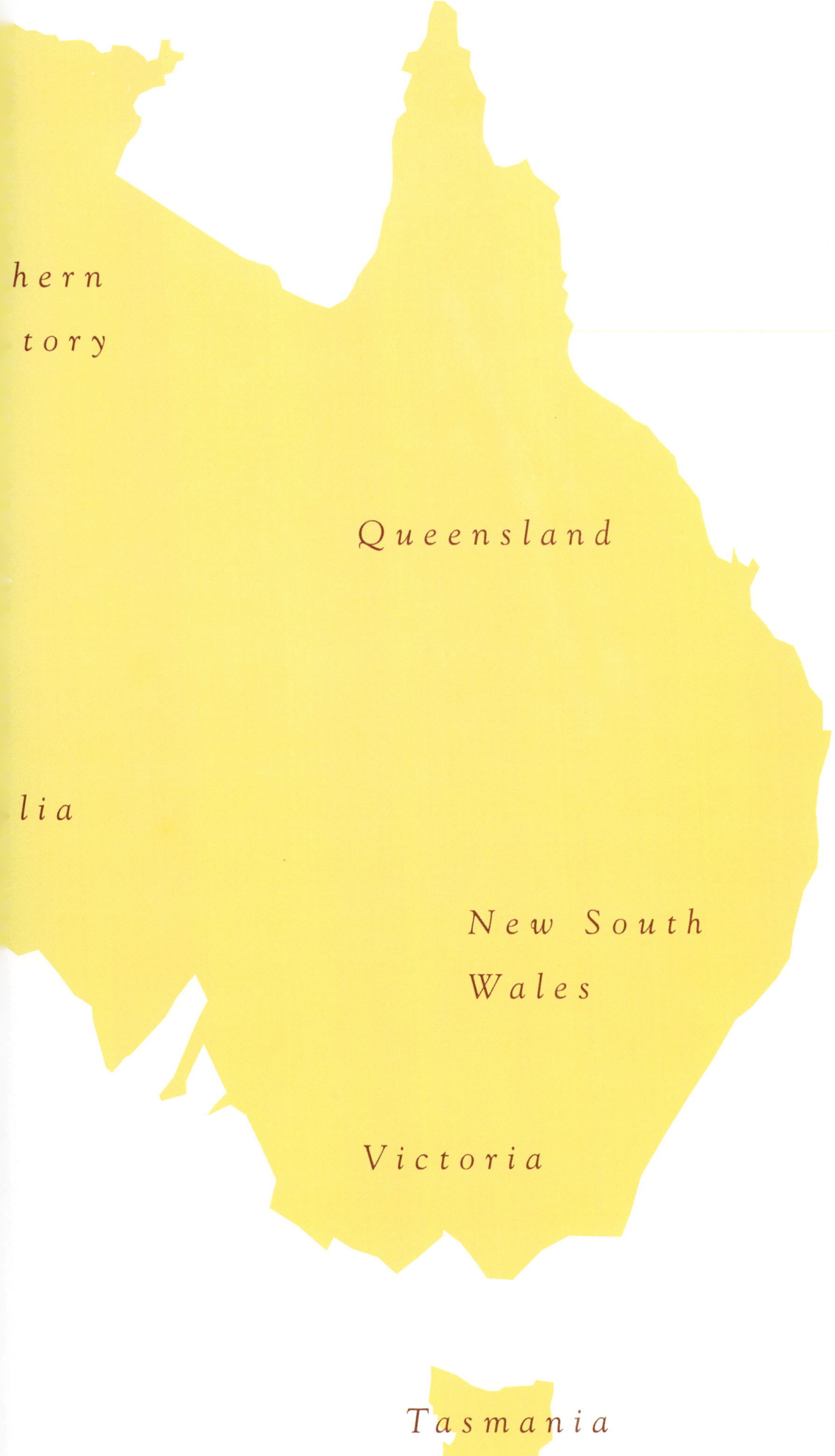
hern
tory
Queensland
lia
New South
Wales
Victoria
Tasmania

CHAPTER SIX

The Meat Industry

An icon of this country's national character, and a multi-billion-dollar income earner, the Australian meat industry is a fascinating tapestry interwoven with history, tradition, imagery, innovation and a restless pursuit of new markets and technology.

Beef cattle and sheep were among the first domestic animals brought to Australia by European settlers, and formed the backbone of early Australian agriculture. Today, pigs and poultry combine with cattle and sheep to provide the vast majority of the meat consumed in this country, as well as a sizeable surplus for export to world markets.

BEEF CATTLE

Apart from the huge cattle stations of the outback, the new face of the beef industry includes a growing element of intensive lot feeding that has emerged in unison with cattle farms in the more closely settled coastal and temperate areas, where production is frequently boosted with grazing crops and improved pasture.

All these elements make up an Australian beef cattle herd that stood at around 25.5 million in mid-1998.

With its relatively small human population, Australia has large surpluses of beef over and above domestic consumption requirements. This makes it one of the world's largest traders of beef.

The 810,000 tonnes of beef Australia will export in 1998 represent only around half of the total beef production in this country. Sales of domestic and export beef, plus live cattle exports, are estimated to earn $5–6 billion.

Throughout its 200 years of commercial activity, the Australian beef industry has seen vast ebbs and flows in its fortunes and cattle numbers. These fluctuations have been driven by domestic and export market prices, and even more so by the vagaries of Australian seasonal conditions.

During 1997 and into 1998, exports of Australian frozen and chilled beef became more reliant on Asia, with Japan (the biggest market), Korea, Taiwan and other regional markets taking around 60 per cent of total exports.

Above: A typical Australian pastoral scene, with healthy cattle drinking pure water from a dam. Australia is world-renowned for its huge outback cattle stations, which have become a part of the national mythology. With a relatively small population and a large cattle industry, Australia is one of the world's largest exporters of beef

Clockwise from above: *Packaged Australian beef being checked for quality and safety before being exported; stacking supermarket shelves with top-quality Australian beef; efficient grain feeding in modern cattle feedlots has provided another dimension in the business of preparing high-quality meat, although the vast majority of Australian cattle are still fattened on pastures; a farm worker drenching a cow to ensure that it is kept in excellent health*

Japan dominated the purchases, with shipments there projected to reach 325,000 tonnes for 1998, a rise from the 312,000 tonnes shipped there in 1997.

The United States market for processing (hamburger) beef took around 250,000 tonnes in 1998. At these levels, it is still a significant importer of Australian beef, although this is less than the almost 400,000 tonnes which the United States took at the peak of its buying in the 1970s and 1980s.

Today's national cattle herd is a multicoloured mix of more than 55 individual breeds and named crosses, represented by no fewer than 32 different breeding societies and associations. This is a far cry from the genesis of the herd, a motley group of black Africander cattle that came to our shores with the First Fleet in 1788, after having been picked up at the Cape of Good Hope in Africa.

With increased white settlement in Australia, the flow of cattle, mainly from the British Isles, continued to increase. By the start of the 1900s, the British Shorthorn breed dominated the Australian herd, with smaller numbers throughout the country of other British cattle, including Herefords, Devons and black Angus.

In the mid-1930s, a group of far-sighted breeders imported a collection of Indian-derived Zebu cattle, which are better suited to Australian conditions. This was followed, in the mid-1950s, by further imports of Brahmans, a breed derived in the United States from Zebu foundations, and Santa Gertrudis, another American breed developed by blending three-eighths Brahman and five-eighths of a British breed. Brahman cattle and their crosses are referred to as *bos indicus*, identifying them with their Indian origins.

Other *bos indicus* cattle to arrive in Australia in smaller numbers were shiwals from Pakistan, red sindhis from the Indian subcontinent and red Africanders from Africa. All these *bos indicus* cattle substantially changed the mix of the northern Australian cattle herd, allowing greater productivity and cattle survival in the harsh environment of the tropics, and welcome resistance to blood-sucking ticks and buffalo fly. As a result, a high percentage of the subtropical and tropical herd now comprises Brahmans and other *bos indicus* types.

Since 1982, there has been further importation of new-generation European cattle, including the French Charolais, Limousin and Maine Anjou, the German Simmental and the Italian Chianina, Romagnola, and Piedmontese. These cattle have brought with them the new benefits of fast growth and heavier muscling – resulting in increased meat yield and, therefore, cash profits, especially when they are cross-bred with established British and *bos indicus* herds.

The Fresh Food People

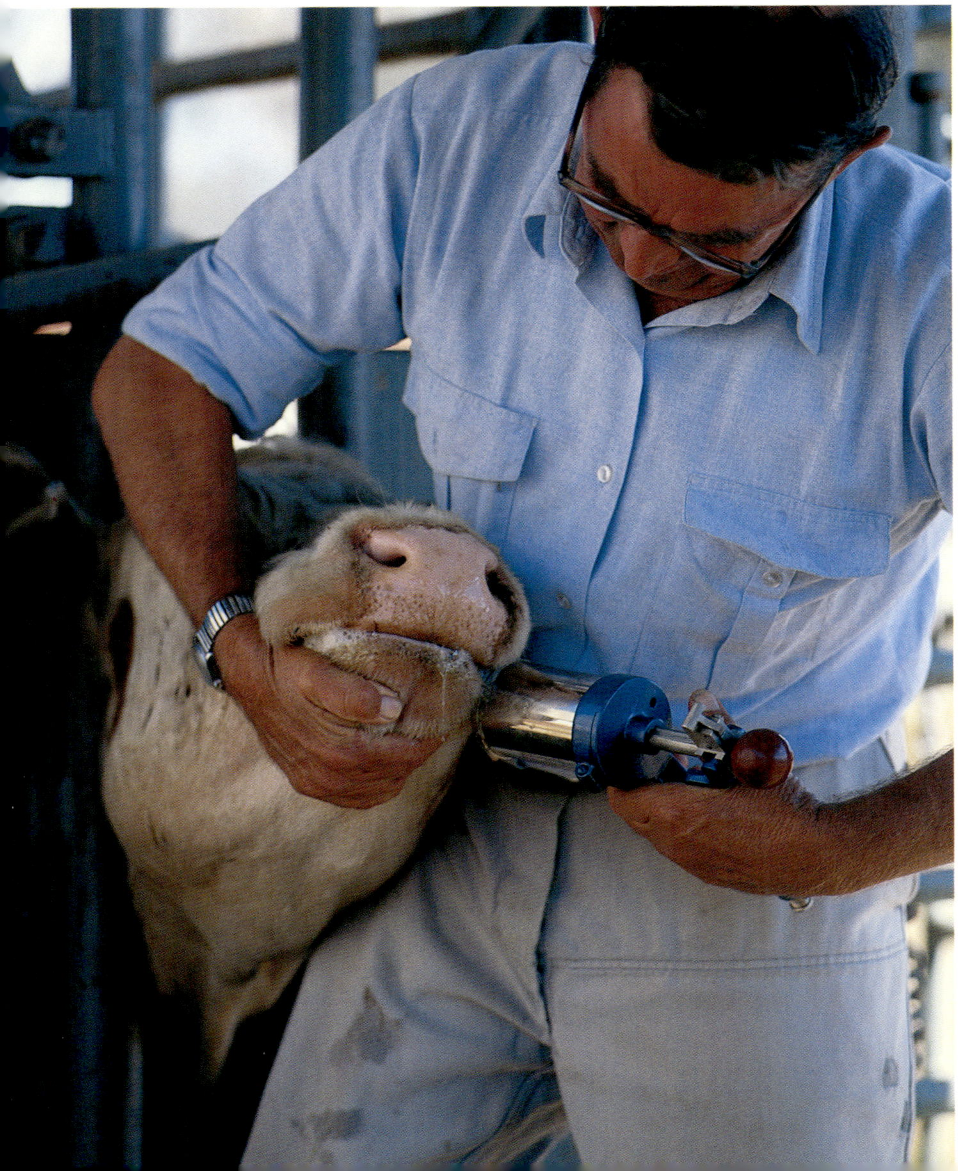

It is from this broad base that the Australian beef cattle herd is now well poised for a broad-based marketing push, aimed at selling beef to domestic and export outlets, as well as live cattle for slaughter and breeding to overseas customers.

Driving industry progress is a blend of huge company producers and smaller owner-operator individuals, all highly efficient and closely attuned to customer needs.

The modern generation of beef producers tend to have excellent environmental and performance records, like the Australian-owned Stanbroke Pastoral Company. Stanbroke is the nation's biggest beef producer, with an archipelago of strategically placed properties, so that cattle can be bred in one region and moved to another for processing, and stocking pressure can be relieved from areas in severe drought.

Major developments are also happening at the customer end of the chain. For instance, new systems such as Meat Standards Australia have been set up to identify superior eating beef more closely that diners can trust and buy again and again.

On the eve of the new millennium, Australia's beef and cattle industry is alive and well poised to continue its reign as the pride of the rural sector.

Clockwise from opposite: *A worker carefully inspecting beef carcases stored in a cold room; Draughtmaster, a breed of cattle farmed in Australia; a Belted Galloway bull; a herd of Brahmans – a breed of cattle derived in the United States from the Indian-derived Zebu – which is used widely in Australia for crossbreeding*

SHEEP MEAT

Sheep meats consumed in Australia come from two distinctly different but, at times, overlapping sources.

One source is the specialist prime-lamb-producing industry, using British-derived meat sheep. On latest figures, around 1.5 million lambs are slaughtered annually to produce just over 280,000 tonnes of meat.

Australia remains a significant exporter of lamb, with exports for 1988 expected to reach about 75,000 tonnes, the remaining production of 205,000 tonnes being consumed domestically.

The other source of sheep meats is the nation's predominantly Merino wool-producing sheep flock of 120 million, which are slaughtered for their meat as a by-product.

From this source, up to 335,000 tonnes of mutton are produced annually, of which 163,000 tonnes are exported and the remaining 172,000 tonnes are consumed in Australia.

Australia's principal overseas markets for lamb are the United States, Asia, the Middle East and, more recently, a growing outlet in South Africa. Major mutton export markets are the Middle East, Asia and also South Africa.

Top: *A flock of Wiltshire Horn sheep grazing in tranquil countryside*

Middle: *White Suffolk is another variety of sheep bred for its meat in Australia*

Bottom: *Displaying a tray of Australian halal sheep meat, which is capturing new overseas markets with its superior taste*

Opposite: *A flock of Merino sheep following a sheepwalk in the fertile Murray region*

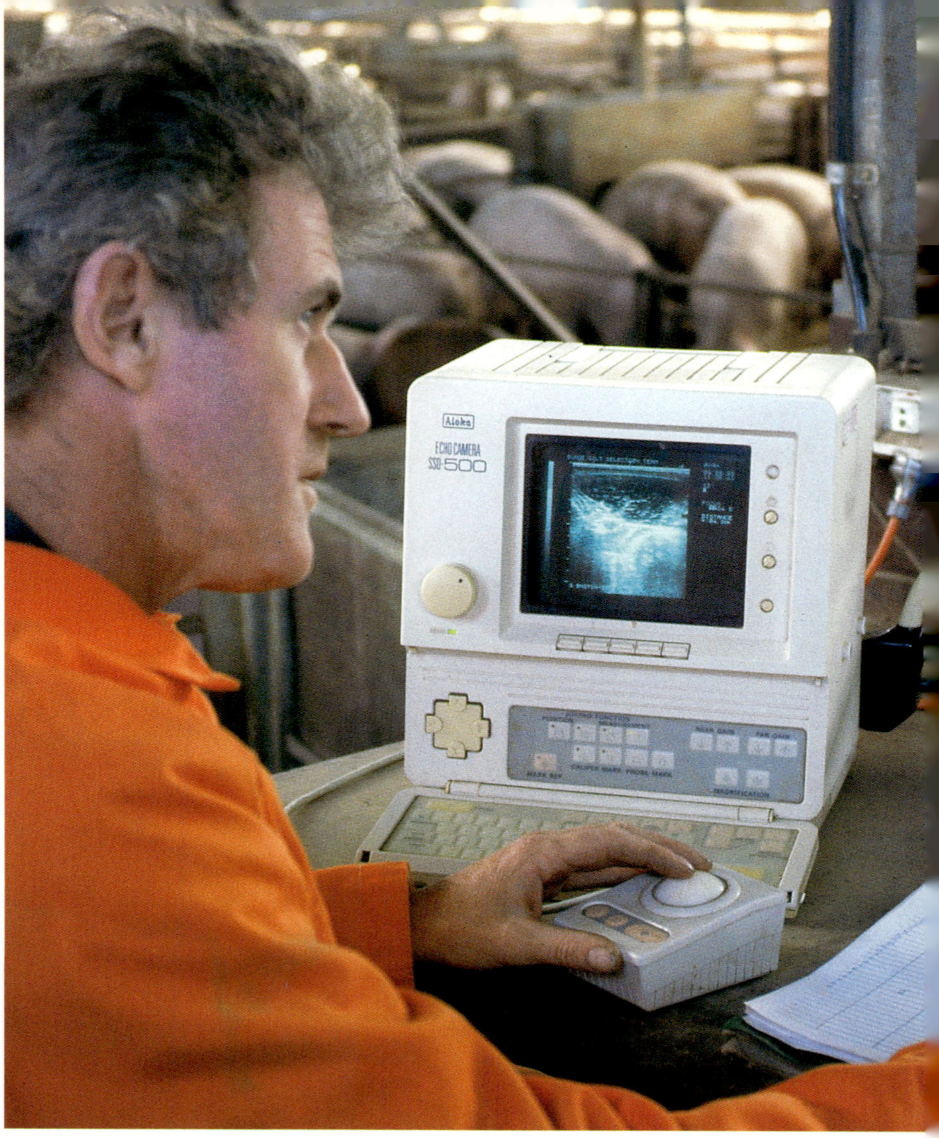
Aloka
ECHO CAMERA
SSD-500

PORK

Pig raising has been part of the Australian farming scene almost from the arrival of the First Fleet.

Up until the mid-1950s, most pig-meat production was linked with the large number of small dairy farms across the nation. Any surplus skim milk was fed to pigs, so that the farms could produce pork and bacon as by-products.

With the change in the dairy industry towards larger and fewer farms producing whole milk, pig production underwent a major structural change. Pigs are now being bred in much larger units and are being fed scientifically formulated grain and concentrate rations. Some of the major piggeries in Australia now have up to 10,000 breeding sows.

Pig raising has been a part of the Australian farming scene almost from the arrival of the First Fleet

Around 3,200 professional pig producers across the nation, breeding about 308,000 sows, account for Australia's total annual output of some 340,000 tonnes of pork and bacon.

Australian pig-meat exports total around 14,500 tonnes, earning the country some $52.5 million, which includes exports of wild boar meat.

Given the increasing sophistication of Australia's pig-meat industry, this is a sector which can be expected to remain a strong export performer for many more years to come.

Opposite (clockwise from top left): *Proper handling and the correct design of handling facilities require sound knowledge of the behaviour of pigs – for example, like other animals, pigs have a tendency to move towards a more brightly lit area; in an intensive piggery such as this, using partially automated systems, it is much easier to maintain hygienic conditions and reduce the risk of diseases; diet formulation involves determining which feeds and their proportions are to be included to provide the desired intake of various nutrients for maximum production*

Above: *a loin of pork seasoned with dates and ginger accompanied by an apricot-glazed ham*

CHICKEN MEAT

Like the pork industry, but about a decade earlier, chicken meat production in Australia underwent a structural metamorphosis.

From a small farm industry, it has become an industry based on very large, specialist producers, often vertically integrated as contract suppliers for major chicken processors, wholesalers and retailers.

The area of poultry raising is distinctly separate from the egg production side of the chicken industry, using completely different types of birds and rations.

Powerful point-of-sale promotion and an enormous technology leap to ensure high quality and complete uniformity in the product have dramatically increased Australia's domestic consumption of chicken meat to around 30 kilograms per capita – a far cry from the days when chicken was a special meal served on Sundays or festive days.

Retail chicken meat sales in Australia are worth around $2.25 billion annually.

A total of 350 million birds slaughtered annually produce around 500,000 tonnes of chicken meat. Although most of this is consumed domestically, chicken product exports are on the increase, driven by a vigorous marketing campaign.

The bulk of present exports are special items like chicken feet to Hong Kong and China, but other lines of value-added chicken are increasing, like cooked chicken products to New Zealand and gourmet wings to other outlets.

Around 12,000 tonnes of chicken product exports earn Australia between $16 million and $20 million.

GAME AND RARE MEATS

A very small percentage of total Australian meat consumption is sourced from farmed rare meats like goat, venison (deer), emu, ostrich, crocodile and kangaroo. Under Australian law, it is forbidden to sell meat from native animals harvested from the wild.

However, as stated earlier, pig meat harvested from the wild is permitted for export, principally to Europe, where it enjoys a virtual cult following. Many game shooters in outback Australia are making a good living from the export of wild pig carcases, prepared under strict protocols of hygiene and refrigeration.

Malcolm McCosker

Top: *A farm worker tends to a batch of freshly hatched chicks*

Middle: *An aerial view of a poultry farm, with characteristically long breeding sheds, set in the Australian countryside*

Bottom: *A game rooster proudly showing off its feathers*

Darling Downs
Bacon Co-operative

Dedicated farmers, skilled staff and more than 90 years of processing experience under a single roof combine to turn out top-selling lines of chilled ham, bacon and smallgoods for the Darling Downs Bacon (DDB) Co-operative.

The Co-operative counts as its members half of the total pig producers in the State of Queensland, and produces not only tried and tested lines of smallgoods, but also popular new varieties aimed at the health-food and export markets.

As well as supplying national supermarket chains and commercial and institutional outlets with its leading and well-accepted range of chilled ham, bacon and smallgoods lines, DDB is now accelerating its production of lean shaved meats, a market which is growing at a consistently strong rate.

Big-name domestic supermarkets including Coles, Woolworths, Franklins and BiLo, as well as independent stores, stock DDB products; however, the Co-operative's market penetration extends deeper than that.

The new shaved products, marketed under the 'Royal Lean' brand name, are set to catch customers' attention. The products are 97 per cent fat free, yet full of flavour.

The range includes a number of ham varieties, as well as pastrami ham and chicken breast meat.

Darling Downs Bacon is also supplying fresh pork to the Franklins supermarket chain.

Formed in 1911 in the rural hub of Toowoomba, high in the Great Dividing Range west of Brisbane, DDB has always been a key member of the community.

As well as enabling farm owners to expand their businesses and maintain staff on the back of a reliable processor, the Co-operative's single processing plant at Willowburn has long been a major job centre for Toowoomba city, employing around 500 hardworking Queenslanders.

Record profits in the late 1990s, burgeoning consumer support and improved production techniques mean that DDB has been ranked among Australia's Top 400 businesses.

Please refer to page 190 for contact details

Top: *90 years of processing history*

Middle and bottom: *Darling Downs Bacon manufactures a wide range of products*

Columbus Line

PRIMARY COOLERS – HOW COLUMBUS LINE CARGO CARE BENEFITS OUR AGRICULTURAL PRODUCE EXPORTERS

The handling of temperature-sensitive cargo such as frozen meat, chilled meat, seafood, citrus fruit or wine requires a seamless supply chain in order to deliver Australia's primary produce into the hands of our export clients' customers as quickly and safely as possible.

Columbus Line is the longest-serving container carrier operating between Australia and North America, having introduced containerisation to the trade in 1971. The company has since established an outstanding reputation for service reliability and Cargo Care.

The transportation of perishable cargo begins with the packing and inland transfer of the cargo to the load port. This is undertaken in specialised refrigerated containers which are designed to maintain temperatures within exacting tolerances set by the United States Department of Agriculture.

Traditional 'reefer' containers are designed for attachment to a clip-on refrigeration unit, which can control the temperature inside the container to within 0.1 degrees Celsius of the set point. On board, the reefers are plugged into the vessel's own Conair refrigeration and computer monitoring system, enabling container temperature levels to be regulated and recorded throughout the journey.

Columbus Line has now developed a new generation of reefers, which incorporate the refrigeration unit within the container itself. These integral containers, or 'integs', allow goods to be cooled to as low as -29 degrees Celsius and microprocessor-maintained to within 0.25 degrees Celsius without being hooked up to the ship's refrigeration system. A laptop computer can be connected for convenient on-shore temperature measurement, control and recording.

Columbus Line's latest innovation is the new 'Maxi-Cube' integrated reefer unit. Built in Germany, these 20- and 40-foot containers offer the largest capacity in the industry. More interior space is achieved through the use of specially formulated CFC-free foam insulation, which allows for a unique slim-line unit with thinner container walls, while maintaining insulation values to ISO standards.

Wine transportation involves another purpose-designed technology which Columbus Line helped pioneer – the thermal blanket. These blankets cover the container walls and roof, protecting the wine against vibration and acting as additional insulation.

While temperature control is critical to product integrity, speed of delivery and flexibility are also of primary importance.

To the West Coast of the United States, Columbus Line offers a schedule of weekly sailings between Melbourne, Sydney, San Francisco and Los Angeles that provides fixed-day cargo availability at all key destinations.

Columbus Line also provides direct services which call every ten days at Houston on the United States' Gulf Coast, and Philadelphia, Norfolk, Savannah and Jacksonville on the East Coast. This 'all-water' service maximises product integrity by minimising double-handling, road and rail transport and inland haulage distances, greatly reducing the risks of pilferage, damage, spoilage or loss.

Columbus Line also services the Pacific Islands, with the company's West Coast North American fleet making fortnightly calls to Suva and Papeete.

Cargo Care, service flexibility, reliable ocean transportation, highly cost efficient inland intermodal connections and a commitment to the development of innovative solutions make Columbus Line 'the logistical choice' for Australia's primary-produce exporters.

Please refer to page 189 for contact details

Opposite top: *Columbus Canada vessel departs Sydney for North America*

Opposite bottom: *A refrigerated container is loaded onto Columbus Victoria in Melbourne for export to North America*

Above left to right: *Columbus Queensland – part of Columbus Line's fleet of container ships servicing the East, West and Gulf Coasts of North America; South Australian wine is transported overland by road or rail to Melbourne for shipment to North America*

Glenmore Export Australia

Top, centre and bottom: *Glenmore's Shanghai Deli, featuring a large range of Australian products*

With its established links with Australia's leading fine-food suppliers and its export experience, Glenmore Export Australia offers an economical means of sourcing high-quality Australian product for small-quantity consumers.

The company consolidates ranges of Australian specialist foods from beef, lamb or veal cuts and local pâtés, to salamis, wines, seafood and dairy produce, hence reducing freight costs and making viable the supply of Australia's gourmet foods to the Asian market.

Servicing international hotels, small supermarkets and providores throughout the Asian region, Glenmore is backed by the long-established Glenmore Meat Company, Sydney's leading meat supplier to the food service industry and retail trade.

Since 1994, when it established operations in Shanghai, Glenmore Export Australia has amassed broad export experience – supplying to markets in Hong Kong, Singapore, Vietnam and Hawaii.

Detailed knowledge of export procedures, documentation requirements, and the certification needed for importing into these and other Asian countries have resulted in a smooth and efficient operation.

Glenmore has access to the complete range of fine foods that Australia produces, as it is not tied to any one line. This means that it can source almost any product to suit the client's needs, whether it be Tasmanian salmon, fresh meat products, camembert cheese, locally brewed beers or Barossa Valley wines.

Freight costs are kept to a minimum by consolidating products into one 1,400-kilogram container in Sydney and air-freighting direct to the client.

Glenmore's suppliers are Australian-owned and take pride in producing high-quality local products – from Hunter Valley wines to Hereford beef cuts, produce is grown or reared in Australia's clean and natural environment.

Glenmore has also been helping international hotels and supermarkets to promote Australian produce by supplying Australian specialties such as whole barramundi, live yabbies, kangaroo, emu and crocodile cuts, and native bush fruits and spices.

Please refer to page 192 for contact details

Stanbroke Pastoral Company

Stanbroke steers destined for the Japanese chilled beef market

Stanbroke Pastoral Company (a subsidiary of AMP Limited) is Australia's, and arguably, the world's largest beef cattle producer.

Based in Brisbane, Queensland, the company owns 30 cattle stations covering 36 million acres across Queensland and the NorthernTerritory, running over 535,000 cattle.

Traditionally, the company has produced grass-fed beef for the North Asian and North American markets, but the emerging markets for grain-fed cattle, live cattle exports and chilled beef into South Asian markets have also been recognised by Stanbroke.

The company has developed its own 'Diamantina' brand for a quality range of processed products for domestic and export markets.

From the lush Central Highlands to the vast natural herbage areas of the Channel Country and the expansive grasslands of the Barkly Tablelands, along with its Darling Downs feedlot, Stanbroke annually markets some 120,000 cattle.

Stanbroke's customers can rely on its product integrity – control, ownership and management of the product, from property to market-ready.

Stanbroke's customers can rely on its product integrity

Please refer to page 201 for contact details

CHAPTER SEVEN

Dairy and Eggs

Thirty years ago, Australia had three million dairy cows and produced just under seven million litres of milk. By 1996–97 the number of cows had fallen by a third, but milk production had risen to a record nine million litres, the fourth consecutive year the record had been broken. Production is forecast to reach 11 million litres by 2002–3. The growth and productivity gains are reflective of an industry which can lay serious claim to being the most dynamic of any in Australian agriculture over the past decade. Over the past five years, it has attracted an estimated $1 billion in new capital investment and exports have grown to represent 40 per cent of production, or nearly $2 billion. Farm productivity has increased by an average five per cent in each year of the 1990s, and milk production costs are among the lowest in the world.

The industry's transformation has coincided with carefully phased deregulation and rationalisation since the mid-1980s, under a series of government–industry plans designed to give it time to adjust to the impact of subsidised global competition. Between 1973, when Britain entered the European Community, ending Australian access to one of its biggest markets, and 1982, milk production fell from 7.5 million litres to 5.2 million litres. However, it has risen steadily ever since, as the unwinding of complex price equalisation arrangements has given dairy companies greater incentive to become more efficient and develop new markets and value-added products.

Above: A selection of washed rind cheeses to tempt even the most discerning palate. Australian cheeses are winning awards all over the world for their delicate flavour and excellent consistency, enhancing Australia's growing international reputation as a producer of fine foods

Australia's dairy industry has traditionally been divided into two distinct sectors:

- 'market' or fresh, liquid milk; and
- manufacturing milk, which is processed into short-shelf-life products, like yoghurt, and less perishable, more exportable products like cheeses, butter and skim milk powder.

Though obviously a similar product, farmers have been paid a significant premium for market milk, to reward them for committing to supply local consumers with fresh milk every day of the year. The rest goes into the manufacturing pipeline.

In Victoria – by far the biggest and lowest-cost dairy State, comprising around two-thirds of Australian production – a very high proportion of output is manufacturing milk, accounting for its status as the country's biggest milk exporter.

The three biggest producers of long-life manufactured dairy products in Australia are Bonlac Foods and Murray Goulburn, both Victorian-based farmer co-operatives, and the Sydney-based Dairy Farmers Group. All of these producers have grown substantially over the past decade.

Deregulation of the fresh milk market has also been occurring gradually since 1986. While prices to farmers remain regulated in all States, the mass of controls which used to exist post-farmgate, including price margins and restrictions on who could sell milk to whom, will have been removed in all States by the end of 1998. This well-flagged process has seen all the major players in the market milk and fresh dairy produce sector jockeying to extend their presence interstate and develop national brands.

Though the possibility of further rationalisation is constantly present, the main processors in the competitive eastern Australian market in 1998 were National Foods, which had developed a strong position in all States except Queensland; Dairy Farmers, which is strong in all eastern States except Victoria; and Parmalat, which is strong in Victoria and Queensland. In Western Australia, where geography has helped to isolate the market, Peters and Browns are the dominant players.

The Australian Dairy Corporation manages the domestic market support scheme, co-ordinates industry-wide marketing and promotion, and controls cheese exports to the premium-paying Japanese market and Australia's quota access to the European Union. It also licenses an industry brand, Dairygood, introduced in the late 1980s to represent clean, fresh Australian dairy produce.

Clockwise from above: *Milk is stored safely in huge storage vats and quickly transported all over the country by an efficient network of suitably equipped trucks; a worker at the Warrnambool Cheese and Butter Factory ensuring that the production lines run smoothly; a Friesian cow – Friesians and Jerseys are two breeds of cattle commonly used for dairy purposes in Australia; the advent of computer technology is revolutionising the dairy industry; cows taking turns in a milking rotation; a Jersey cow grazing in a lush green paddock in Tasmania*

Overleaf: *A farmer feeding free-range chickens at Granville Harbour, on Tasmania's west coast. The free-range egg industry is now estimated to comprise five per cent of Australia's total egg production*

36

CROPLIFT
FERTILISER

EGGS

Like the dairy industry, the egg industry has experienced significant change over the past decade due to the twin forces of market deregulation and changing consumer behaviour.

Ten years ago, all Australian States had statutory boards which set wholesale egg prices and established egg pools. Hen quotas were also introduced throughout the country in the 1970s.

Today, New South Wales, Victoria, Queensland and South Australia are totally deregulated and Tasmania has removed marketing controls, although it is yet to abandon production quotas. Western Australia is the only State still to retain a traditional egg board.

While there is some disagreement over the figures, Australia's egg production is of the order of some 200 million dozen per year...

In New South Wales, Victoria and South Australia, the boards have been replaced by producer co-operatives and family-run businesses. Victoria's co-operative, renamed Farm Pride Foods Ltd, took the further step in 1998 of floating on the Australian Stock Exchange, in the process becoming the first co-operative to convert to a public company and list under new Australian Stock Exchange rules, designed to attract more farmer-controlled businesses to the exchange.

While there is some disagreement over the figures, Australia's egg production is of the order of some 200 million dozen per year – the Australian Bureau of Statistics puts the figure at 170 million dozen, whereas the Australian Egg Industry Association (AEIA) estimates it at 210 million dozen. Of that, 25 million dozen are estimated to be laid in Australian backyards, indicating that the chookyard is still a fixture in many Australian homes. The rest come from an estimated 1,000 egg producers.

The biggest egg producer, Pace Farms, based in Blacktown, near Sydney, grosses approximately $70 million per year from some 1.5 million chickens on 40 poultry farms and from associated value-adding operations.

In total, Australia has a flock of 10 million laying hens, with New South Wales accounting for 39 per cent of those, and Victoria accounting for another 24 per cent. The farmgate value of their output is around $300 million, according to the AEIA, while retail sales are put at $525 million.

In a reflection of changing consumer habits, around 55 per cent of production is estimated to be sold as 'shell eggs' – eggs in their shells – by supermarkets, and

another 15 per cent as shell eggs by other retailers. A further 15 per cent is now accounted for by shell eggs sold to the food services sector, and the remaining 15 per cent is sold in processed form – such products as whole egg powder, fresh egg yolk, fresh whole egg, salted egg yolk and egg white powder – to food manufacturers.

Per capita consumption of eggs in Australia is stable at around 156 eggs per year. However, Farm Pride Foods, which has one of the biggest value-adding egg processing plants in Australia, says that consumption is shifting from shell eggs to processed egg products as part of the global trend among consumers to eat more food away from home and to buy the food which they cook at home in a prepared or semi-prepared state. In the United States, consumption of processed egg products has increased, from 13 per cent of total egg consumption in 1980 to 27 per cent in 1995.

Due to increasing consumer demand, the free-range egg industry is also growing rapidly, and is now estimated at five per cent of Australia's egg production – a doubling in three years. Similarly, barn-laid or deep-litter eggs have grown from nothing to two per cent over the same period.

However, the executive director of the AEIA, Hugh McMaster, does not believe the same commercial pressure will be brought against caged-hen production as has occurred in Western Europe. 'So long as the industry offers a choice, consumers can vote with their feet,' he says.

The dairy and egg industries are among the most dynamic sectors of Australia's agriculture industry. And with a combination of enviable natural resources, stringent quality controls and leading-edge technology, Australia's dairy and egg producers are also world leaders in their respective fields.

Cathy Bolt

Opposite (top to bottom): *A scientist observing the characteristics of egg shells after the introduction of a new feeding program; an exterior view of a poultry layer farm, showing a single silo for premixed feed*

Top and above: *These hens are reared to lay eggs, which are automatically collected using a sloping metal rack attached to the cages. The cages are also designed to allow automatic manure removal*

Warrnambool Cheese and Butter Factory

Top: *The Warrnambool Cheese and Butter Factory has developed a range of dairy products tailored to the needs of customers worldwide*

Centre left: *The company prides itself on its paddock-to-plate quality philosophy*

Centre right: *Product being loaded for export*

Bottom: *The main chamber of the new Niro Milk dryer commissioned in 1998*

In the south-west of Victoria, the Warrnambool Cheese and Butter Factory is developing new and innovative products tailored to meet the tastes and needs of emerging world markets. Established in 1888, the company is moving its 100-year focus from bulk commodities to value-added products and finding significant success in exporting.

The Warrnambool Cheese and Butter Factory exports more than 60 per cent of its products, which range from fresh, low-fat and flavoured milks to butters and creams, dairy blends, whey protein concentrates, bulk cheeses and milk powders. Key international markets include the United States, Japan, Korea, the Philippines, Russia, the Middle East and Africa.

Developing a close liaison with each customer and creating products that meet specific requirements form the core of business relationships for the Warrnambool Cheese and Butter Factory. With the Westernisation of Asian dietary habits, dairy products are being tailored to meet a growing demand. Milk powders that are calcium fortified and butter products that have the composition of cheese are just some of the products made to order for overseas markets.

In the relative isolation of country Victoria, the Warrnambool Cheese and Butter Factory takes advantage of the clean environment to make products free from heavy metals, harmful chemicals and radioactive fallout. The company has been given a clean bill of health from the Australian Dairy Authority's national chemical residue survey, and all manufactured dairy products are tested at an on-site laboratory, accredited by the National Association of Testing Authority.

From its beginnings with just six farmers, the Warrnambool Cheese and Butter Factory has grown to more than 600 suppliers and is turning over more than $150 million a year. A sustained expansion over the past decade has increased capacity fivefold and has given this long-established company the modernised plant and equipment required for growing demands. Beginning with the highest quality raw materials, the Warrnambool Cheese and Butter Factory is developing healthy, flavoursome products delivered fresh in the quickest possible time.

Please refer to page 204 for contact details

Pace Farms

KEEPING THE PACE ON THE FARM

Great egg products are found in the healthy environment encompassing Australia's egg industries. For 50 years, Pace Farms has played an integral role in this industry, becoming the largest independent producer and marketer of quality eggs, egg products and specialty egg items in Australia today.

Today, Pace Farms supplies McDonald's Australia, as well as the major supermarket chains of Woolworths, Coles and Franklins.

Pace's stocks of two million birds are housed at locations throughout Australia, management implementing differing farming practices according to demand and the tastes of particular market segments. The expertise Pace Farms maintains ensures a consistent, guaranteed supply of premium product.

From day-old chicks at Pace Farms' rearing facilities to egg-producing hens, every aspect of production and maintenance is strictly applied in line with the rigorous standards of Pace Farms' personally tailored HACCP food safety program.

Strict bio-security practices are in place on all Pace Farms' properties to protect the health of flocks. Pace's consultant veterinarian ensures expert practice, complementing its comprehensive in-house vaccination program. Feed is formulated by skilled avian nutritionists, who pick only choice grains and natural ingredients containing no artificial additives or colourings.

Birds are housed in accordance with the Model Code of Practice for the Welfare of Animals, prepared by the Agriculture and Resources Management Council of Australia. The Code sets out standards for flock housing, feed, water, lighting, ventilation, health, space allowances, flock protection, temperature control and management practices.

Once the safest, cleanest, highest quality produce is ready, Pace Farms' modern fleet of refrigerated transport vehicles delivers peak condition product rapidly to its customers.

Pace Farms' products are acknowledged by industry watchdogs such as the Australian Heart Foundation and the Halal Authority of Australia, with export registration approved by Australian Quarantine Inspection Services. Committed to retaining leadership of the Australian egg industry, Pace Farms is supremely positioned for quality supply into the future.

Please refer to page 198 for contact details

Above: Our RSPCA endorsed Liberty® egg facility

Below: One of many of Pace Farms' fleet of modern refrigerated vehicles

Bottom: Pace Farms' state-of-the-art egg grading factory located on the outskirts of Sydney

CHAPTER EIGHT

Seafood

With 36,000 kilometres of coastline and the world's third-largest fishing zone, nature has blessed Australia with a substantial seafood reservoir, and the Australian fishing industry is making good use of this advantage. Seafood production ranks as the country's fifth most valuable rural industry after wool, beef, wheat and dairy products. Every year, fishing provides around five per cent of Australia's rural exports, with the total value of production exceeding $1.7 billion and providing direct employment for close to 20,000 people.

One of the fastest growing segments within this valuable industry is the science of aquaculture, which has created the relatively new activity of 'farming' high-value, high-quality fish such as salmon, trout and tuna. Heavy investment in this activity is supplementing traditional fishing output and setting the scene for strong future growth of seafood production.

Fishing has been part of Australia's history from its earliest days. Australia's Aboriginal people were the first fishermen, working the shallow offshore waters and inland rivers. European settlers were quick to follow suit after their arrival in 1788. Whaling and sealing actually ranked as Australia's biggest early industries. Whale oil, whale meat, whalebone and sealskins were the country's biggest export items until overtaken by wool in 1834. The last whales taken were in 1978, at the Cheynes Beach station near Albany on Western Australia's south coast, two years before an outright ban on all whaling.

From their start in whaling, Australia's early fishermen branched out to catch some of the estimated 3,600 species of fish which inhabit local waters. They soon found that few of this astonishing array have genuine commercial value. Fewer than 100 species are currently fished commercially.

Today, the most valuable seafood varieties caught by the Australian fishing industry are rock lobster, prawns, tuna, abalone, scallops and a wide variety of fin fish, from shark to orange roughey. Pearling, although technically not in the seafood category, is another big

Above: Fishermen at work on Australia's pristine coastal waters. Australia has the world's third-largest fishing zone, and fishing has been an intergral part of the country's economy from its earliest days, with the Aboriginal peoples being the first fishermen. Today, fishing provides around five per cent of Australia's rural exports annually and the fishing industry employs about 20,000 people directly

marine-based industry, and many consider oysters to be one of the greatest seafood delicacies.

According to the Bureau of Agricultural and Resource Economics, lobster, mainly from the west coast, represents about 24 per cent of the national fisheries industry's value. The United States and Japan are currently the largest export markets for Australian lobsters, with the premium-quality Australian product commanding very high restaurant prices worldwide.

Prawns are the second most valuable seafood product, with an annual value of around $329 million, or about 19 per cent of the nation's total catch. The value of the annual abalone harvest is around $148 million (eight per cent of national seafood output), while tuna accounts for about $115 million (6.5 per cent), scallops $51 million (three per cent), oysters $42 million (2.5 per cent), and other fin fish $363 million (21 per cent).

While the figures look good, some are of the opinion that there remains a vast potential for increased output. One international report by the United Nations agency, the Food and Agricultural Organisation (FAO), estimated in 1970 that Australia's waters could yield an annual catch of 2.3 million tonnes. At present, the catch is only ten per cent of this estimate.

...the United Nations agency, the Food and Agricultural Organisation (FAO), estimated in 1970 that Australia's waters could yield an annual catch of 2.3 million tonnes

Other observers of the seafood industry have voiced their concern that to push for a large annual catch, such as that suggested by the FAO, might jeopardise the delicate balance in some of Australia's fisheries. Despite a relatively low annual catch by the standards of other countries, Australia is keenly aware of the dangers of overfishing. Strict management rules now govern much of what happens in the fisheries zones, and quotas are applied to most species harvested directly from the sea – increasing the attractiveness of constructed fishing industries, such as salmon and trout farming.

Opposite: *With 36,000 kilometres of coastline, Australia enjoys a natural advantage in the fishing industry. There are approximately 17 designated fisheries around Australia's coast*

Top and above: *Scallops and orange roughey, a popular fin fish, are among the most valuable varieties of seafood caught by the Australian fishing industry. Other valuable varieties include rock lobster, prawns, tuna and abalone*

Overleaf: *Two fishermen hauling in the catch of the day. Australia is keenly aware of the dangers of overfishing, and has strict rules governing fishing practices and quotas*

GLORIOUS FOOD
FRESH NSW SEAFOOD

TRANSVAC

Australia has approximately 17 designated fisheries around its coast, ranging from the southern shark fishery off the coast of Victoria, to the western deepwater trawl, the northern prawn fishery and the east coast yellowfin tuna fishery. The fisheries occupy most of Australia's nine million square kilometres of the exclusive fishing zone which extends for 200 nautical miles off its coastline and incorporates the seas around remote Australian outposts such as Heard, McDonald and Kerguelen Islands. In simple area terms, the Australian Fishing Zone is 16 per cent larger than the country's land mass.

Over the years, the Australian fishing industry has progressively moved to harvesting new species. Coastal resources, including the luscious Sydney rock oyster, were among the earlier shellfish to be exploited. Crabs, prawns and lobster soon followed. By early this century, fishermen had begun to move off the New South Wales coast to catch barracouta, with big catches being reported from 1911 onwards. In the 1920s, school shark was being caught off the south coast, along with salmon and southern bluefin tuna.

The rock lobster industry began its rapid rise to the top of the export table in 1947, when the first shipments made their way to the United States and Europe. In the postwar years, as fishing boats and the catching technology improved, there was a dramatic upswing in total production. Prawn trawling started off New South Wales, moved north to Queensland, and then across to Western Australia and into the Northern Territory. Today, the northern prawn fishing zone is considered to be fully exploited. In the 10 years to 1990, the volume of fish landed by Australian fishermen grew by more than 200 per cent, with the number of active fishing boats swelling to an estimated 10,000.

While traditional fishing practices are well established in Australia, aquaculture is just beginning. In 1990, the total value of aquaculture was around $200 million, or close to 20 per cent of the total catch. Salmon, trout and oysters dominate this form of controlled seafood farming. Tuna is also being farmed in controlled conditions off the South Australian coast. Increasing quantities of prawns and freshwater lobster are also being farmed for the international market. At this rate, the growth of aquaculture to the point where it will be the major source of Australian seafood seems inevitable.

As Australia's seafood industry moves from traditional fishing practices to more controlled farming, there is no doubt that its success will continue into the new millennium. The enormous area of coastline and open sea available for fish farming presents a superb opportunity for new investment, especially as technology and science find solutions to enhance yields and boost production.

Tim Treadgold

Top: A selection of Japanese sushi made from premium Australian seafood

Above: A selection of smoked fish products, popular in both local and overseas markets

Opposite (clockwise from top left): Buyers choose yellowfin tuna at the Sydney Fish Market's Sashimi Pavilion while a seller looks on; a wide range of fresh New South Wales seafood is proudly presented by staff at one of the many retail outlets where it can be found, both in Australia and overseas; an aquaculture farm run by the Tasmanian seafood company Tassal, which is a leading producer of farmed Tasmanian salmon

Tasmanian Seafood

Aquaculture is often described as the way of the future for Tasmania's fishing industry. Shining out against a gloomy background of over-fishing of wild fisheries around the world are the bright lights of Tasmania's rapidly developing fish farms.

The State's wild fisheries – particularly for much sought-after abalone, rock lobster and scallops – are carefully monitored and have reached the limits of sustainability, explaining the premium prices received for limited production.

But from a fledgling and largely experimental industry just over 10 years ago, aquaculture has thrust Tasmania onto the world stage as a high-quality producer and has grown to become an economically vital, multi-million-dollar export industry.

The State's natural advantages of being pollution-free and 'clean and green' are more than marketing buzz-phrases, but are major advantages for fish and shellfish producers competing on world markets.

Tasmania is one of the last remaining wilderness areas on earth, boasting crystal-clear waters and the cleanest air on the globe.

The unpolluted waters of the Southern Ocean are the ideal temperature for salmon production and have a naturally high oxygen content – a combination of factors which has seen the State's leading producers, Tassal and Nortas, become key players on the world scene. These quality producers have put Tasmanian salmon products at the forefront of premium markets, where only the finest produce is sought.

Farmed oysters and mussels grown in the State's pristine waterways have found favour on markets around the world for their flavour, quality and freedom from residues.

In terms of economic value, marine farming is expected to far outstrip returns from wild fisheries in the near future – good news not only for the island State, but also for seafood lovers around the world.

Nortas

Above: *Smoked salmon sides on a rack, just out of the smokehouse*

Nortas salmon and salmon trout start their life at the company's hatchery near the Mt Field National Park, on the edge of Tasmania's south-western wilderness, where crystal-clear mountain water is drawn from the Tyenna River.

Young salmon and trout are transferred to the pristine sea at three sites off Bruny Island, where they are held in cages and hand-fed for a further 12 months.

The majority of Nortas fish are chilled and packed on ice for their journey to destinations within Australia and South-East Asia. Shipment to markets within 24 hours of harvest ensures the highest quality, and makes Nortas salmon and salmon trout among the freshest available.

The remainder of the fish go to the company's value-adding factory in Hobart to be processed as a range of premium-quality, high-value, hot and cold smoked salmon products.

Nortas was established just 12 years ago, but has grown rapidly to become a major force in the Tasmanian aquaculture industry.

The privately owned Tasmanian company now employs more than 100 people at its hatchery, grow-out farms, marine farms and value-adding factory. Nortas boasts sales of more than $20 million a year, and has an enviable international reputation for quality.

Please refer to page 196 for contact details

Tassal

Multi-award-winning seafood producer Tassal has set the standard for the seafood processing and aquaculture industries in Australia and around the world.

Tassal's Royal Tasmanian salmon are farmed in the cool, clear waters of southern Tasmania – a pristine environment where the company produces what it confidently calls seafood of unparallelled quality.

This is an opinion shared by chefs and gourmets around the world, where markets have been established for Tassal salmon – in recognition of its well-coloured, firm flesh and delicate taste.

The company's systems, hygiene, safety, and production methods have also won international recognition through ISO accreditation, an achievement which sets Tassal at the forefront of the industry.

As well as establishing their own branch in Japan, Tassal has introduced a distribution system which enables their seafood to arrive in markets in Japan and other Asia–Pacific destinations within 24 to 48 hours.

Tassal's annual production volume is more than 4,500 tonnes of seafood, making the publicly listed company the largest salmon aquaculture company in Australia, and a world leader for innovation and quality production.

Above: *Fresh whole salmon inspection*

Please refer to page 202 for contact details

CHAPTER NINE

Cereals and Grains

Central Sydney is not the place most people picture as the home of Australia's cereals industry. But not long after the first fleet of European settlers arrived in 1788, this was the spot chosen to sow a crop of wheat; in those days, bread was truly 'the staff of life'. It would take a few more years for the settlers to master Australianconditions. Once this happened, however, Australia was launched as a premium supplier of quality grains to the world.

Today, Australian wheat, barley, rice, oats, sorghum and maize, plus a range of other exotic grains, find their way to the tables of families in more than 50 countries around the world. From its humble beginnings, the Australian cereals crop has grown to its current worth of around $4 billion per year, with most of it being exported.

Wheat dominates the cereals business in Australia, accounting for around 60 per cent of the industry. Barley represents around 18 per cent of the crop, with sorghum, rice and oats coming in at between four and seven per cent each. Exact figures vary with the season and market demand. The weather, in particular, can make or break a crop, as shown dramatically in the early 1980s. In 1983, the wheat crop crashed to just nine million tonnes under the impact of drought. The next year, when the rains came, Australia's farms bloomed with a 22-million-tonne crop.

In world terms, the Australian wheat crop represents just three per cent of total production. However, a relatively small domestic market means that around 80 per cent of total production is available for sale to overseas customers. This high export level means that Australia ranks among the world's top four cereal and grain exporters, alongside the United States, Canada and the European Community. Demand for Australian wheat is strong, with the country maintaining an excellent reputation for delivering a premium product which is pest- and disease-free, and low in insecticide residue.

Every year, more than 40,000 farms, mostly family-run, participate in the annual ritual of harvesting, followed by the long haul on trucks and trains to export terminals around the coast. Payment for wheat sales

Above: *Rice is an increasingly popular grain crop, exported to many overseas countries. Rice is the staple food in many of Australia's neighbouring countries in Asia, forming a huge potential export market which is already being explored by many local producers. Australia ranks among the world's top four cereal and grain exporters. Australian cereal and grain exports enjoy excellent reputations for being free of pests and diseases, and low in insecticide residue* remains one of the pillars of the national economy, especially in rural and regional Australia. Selling the crop is the job of State-based grain handling authorities and AWB Limited, the national grains marketer, which is in the process of major changes to meet the challenges of global competition and deregulation.

Barley, Australia's second-ranking cereal, has enjoyed remarkable growth over the past 50 years, with the biggest demand being for malting barley, a critical ingredient in beer production. Many of Japan's biggest-selling beers are made from Australian barley. Fast-growing, hardy and adaptable to a wide range of soil types, barley fits well into crop rotation cycles.

Oat is also popular as a secondary crop, with its biggest market being stockfeed. A small proportion is also marketed for human consumption.

Sorghum remains popular in Queensland, while minor cereals such as millet, maize, rye and the hybrid triticale (a cross between wheat and rye) are proving useful in areas of very light soil.

Cereals have been among Australia's top ten exports for the past century. But getting to that point has been a challenge

Cereals have been among Australia's top ten exports for the past century. But getting to that point has been a challenge. Growers have to be ever-vigilant about the possibility of drought, flood and pest invasions. Quite apart from the vagaries of nature, local growers also face stiff competition from subsidised foreign exports, which hold prices down and destabilise world markets.

Above: *Scientific and technological advancements have played major roles in the successful cultivation and processing of cereals and grains in Australia*
Opposite (clockwise from top left): *A selection of the cereals and grains harvested in Australia today – lupin is a herb of European origin with edible seeds cultivated from ancient times; sorghum is a cereal grass of many varieties which may be divided into four main groups – sweet sorghums, grain sorghums, grass sorghums and broomcorns; triticale, a hybrid wheat and rye cereal; corn – the edible grains, or maize, are embedded in the corncob*

For almost 60 years after that initial central Sydney planting, Australia produced only sufficient cereals for the local market. During this time, the country was a net importer of grain, as farmers faced the challenges posed by labor shortages, imported seeds and growing conditions which were very different from those they had been accustomed to in Europe. It was not until 1845 that wheat exports began on an irregular basis. However, the farmers persevered in their efforts, and finally, in about 1870, regular shipments of Australian grains started making their way to overseas markets.

In the 1870s, Australia's farms struggled with yields of less than half a tonne of wheat per hectare. Today, the average yield often exceeds 1.5 tonnes of wheat per hectare. Science has played a major role in this agricultural success story, through many years of extensive research and development.

As knowledge of the seasons and the land improved, Australia's cereals industry boomed with 'land rushes' in all of the mainland States, as farmers opened new areas to grain farming.

The boom years of the cereals industry in Australia coincided with the industrial revolution. This had a double-barrelled effect on Australia. New markets were being created for grain, as workers moved off the small farms of Europe and into factories. Those same factories were producing remarkable new machines to work the soil and transport the harvest.

Railways were critical in this changed environment, with low-cost transport opening up vast new areas of inland Australia, helping remote farmers overcome the 'tyranny of distance' and providing the mechanism to deliver their crops.

A 'crescent' of land stretching down the west coast and up from South Australia through western Victoria and New South Wales quickly became known as 'the wheat belt'. During the 1860s, the wheat-growing area doubled, and in the following decade it tripled. Most States saw a boom in wheat production, starting in South Australia, which earned an early reputation as the 'granary to the nation'.

New land-clearing techniques, improved use of machinery, railways to haul the crop, and the use of superphosphate fertiliser transformed Australian farming.

Above: *Wheat being processed under the warm Australian sun. Wheat is the major cereal grown in Australia, accounting for around 60 per cent of the industry. Although the Australian wheat crop represents only about three per cent of world production, a relatively small domestic market means that around 80 per cent of local production is available for export. A 'crescent' of land stretching down the west coast and up from South Australia through western Victoria and New South Wales is known as Australia's 'wheat belt'*

Opposite: *Oat harvesting is made easy with state-of-the-art machinery. Oat is a popular secondary crop, suitable for human consumption and for use as animal feed. New land-clearing techniques, improved use of machinery, railways to haul the crop, and the use of superphosphate fertiliser transformed Australian farming into the highly efficient, internationally competitive industry that exists today*

Innovation was also an Australian cereals-industry trademark. Unique machinery was devised to handle Australia's virgin countryside. The 'stump-jump' plough enabled the quick transformation of bush into productive farmland. The 'stripper' speeded harvesting, and H V McKay's invention of the 'combine harvester' pooled together into one machine the jobs of stripping grain from the stalks, threshing and cleaning the wheat, making it ready to be bagged. The 'Sunshine' harvester of 1895 remains one of the great Australian inventions.

Science, however, has been the real key in the growth of Australia's cereals industry, especially in understanding the chemistry of soil and in breeding cereal varieties for Australian conditions. Nitrogen and phosphorous levels were particular challenges which attracted attention as early as 1879, when experimenters began using various fertilisers. By the early 20th century, an application of superphosphate with each planting was a standard procedure.

Seed breeding has also been a critical development, highlighted by the work of William Farrer, an English-born mathematics graduate who migrated to Australia in 1870. From a small farm at Queanbeyan, near Canberra, he experimented with wheat varieties, seeking a strain which would grow strongly in a dry climate, resist the scourge of rust disease and produce a high-quality grain. His discovery of the 'Federation' variety earned him national fame and, for a time, pride of place on Australia's now-defunct $2 note.

MASSEY FERGUSON
587
SUNSHINE
MF
FORD

Breeding the right seed to particular conditions and market demand has become a hallmark of Australian cereals, with names such as Bobs, Cedar, Comeback, Cadoux and Nabawa evolving alongside the famous Federation grain. One of the most recent changes in this area is the practice of growing wheat varieties which can be used to produce high-quality noodles. The same person buying an Australian malt beer in Tokyo today may later enjoy a meal of premium Australian wheat noodles.

Whatever evolves in the marketing process, it seems certain that Australia's farmers will retain their position amongst the world's major suppliers of quality cereals

A feature of all Australian wheat is its white colour, with the once-traditional red-grained wheat being discouraged in order to create a uniform national product. Purity in colour is matched with consistent hardness and quality in starch and protein levels.

Meeting the demands of the market, rather than just growing large crops, is the major trend in Australian cereal production today. Having overcome natural adversity, found methods to improve soil quality, and bred seeds specifically for Australian conditions, the challenge now is to tailor production to what the customer wants. This is leading to many changes in the way Australian grains are produced and sold.

But whatever evolves in the marketing process, it seems certain that Australia's farmers will retain their position amongst the world's major suppliers of quality cereals. Each year, more than 12 million hectares of grains are cropped, and Australia produces anywhere from 15 million to 25 million tonnes of assorted cereals – bearing testament to a world-class industry producing a world-class product.

Tim Treadgold

Opposite: *Night harvesting of wheat in Queensland. Wheat harvesting is commonly done at night, partly due to the cooler temperatures, which allow the produce to remain fresh. A feature of all Australian wheat is its white colour, with the once-traditional red-grained wheat being discouraged in order to create a uniform national product. This purity in colour is matched by the quality of the wheat*

Above: *Australian cereals being loaded for shipment to overseas destinations. Australia's position as one of the world's top-quality cereal producers is well established. For example, many of Japan's biggest-selling beers are made from Australian barley, and noodles made from Australian wheat can be found in many countries around the world*

AWB Limited

AWB Limited is the Australian grains marketer. Its vision is to be Australia's leading agribusiness, building wealth for customers, owners and the nation.

As Australia's leading agricultural exporter, AWB Limited markets 'quality grain made to order' to more than 100 customers in 50 countries around the globe on behalf of its grower shareholders. An Australian brand of wheat has been developed for virtually every wheat flour use around the world, and more than 50 different products are offered to customers.

This marketing effort is supported by technical assistance to customers to add value to their products, and by a vigorous, ongoing research and development program designed to ensure that Australia's growers respond rapidly to customers' changing quality requirements. The white-grained characteristic of Australian wheat, together with stringent quality control, have given Australia an excellent reputation in the international market for clean, dry, white-grained product.

These factors, and AWB Limited's single-desk, integrated system of marketing, ensure that Australia's 45,000 wheat growers receive premium prices for their grain in open competition on the world market.

Australia is a small wheat producer by world standards – accounting for about three per cent of world production – yet ranks fourth of the five major exporting countries that together account for 90 per cent of world trade. Australia's annual wheat exports, comprising around 80 per cent of the total crop, amount to about 10 to 12 per cent of inter-national trade in wheat.

Australia's integrated marketing system is funded by growers, and AWB Limited is making the transition from statutory authority to Australia's largest grain agribusiness that is owned and controlled by growers. This integrated marketing system has allowed economies of scale in research and development finance, marketing, promotion and relationship building, while delivering cost savings through co-ordinated logistics.

AWB Limited is also a major marketer of pulses, oilseeds, coarse grains and other feed grains to domestic and international markets.

Top: *Delicious fresh bread made from top-quality Australian wheat*

Centre: *Grain storage facilities are located strategically in wheat-growing areas*

Above: *Wheat crop ready to harvest under the Australian sunshine*

Please refer to page 186 for contact details

The Lentil Company

Pakistanis devour them in curries. Turks savour them in soups. In breads, cakes, baby foods, snacks, sprouts and much more, the premium-quality Australian lentil is now finding its way into the diets of more than 30 countries around world. The extraordinary growth of the export market is largely due to The Lentil Company (TLC). Founded in 1992 by two visionary Victorian farmers, TLC's annual sales have grown from $600,000 a year to more than $15 million. The key has been to produce a high-quality product for niche markets and ensure reliability in a notoriously uncertain industry.

With a strong research and development base and a commitment to up-to-date grower education, TLC has nurtured the fledging lentil industry through a period of remarkable growth. When the company began, Australia was importing lentils for its own consumption. Now, more than 550 TLC growers plant in excess of 40,000 hectares, destined for both domestic and overseas markets.

Personal contact with growers ensures each crop is given plenty of 'Tender, Loving Care', guaranteeing the lentils are pure and fresh when delivered. The superior quality is tempting export markets back for more each year. TLC has turned three lentil varieties, the Cobber, Digger and Matilda, into 20 different product forms, including quality-assured lentils from the farm, giving it the flexibility to meet the needs of markets around the world. Even major lentil-producing countries such as Turkey, Canada, New Zealand and the United States are now importing the Australian TLC products.

TLC exports 95 per cent of its production. All lentils are processed under an Australian Quarantine Inspection Service–accredited program, and are tailored to individual requirements.

TLC has been recognised with a string of awards for export innovation. As well as expanding its global lentil markets, it is now applying its winning formula to other specialty crops – from chickpeas to broad beans, faba beans, fenugreek, peas and canary seed.

Top to bottom: *TLC co-founder, Peter Blair, tends his lentil crop; TLC use modern harvesting, transport and processing equipment; when you see the TLC bag… think quality!; a selection of TLC's award-winning lentils*

Please refer to page 202 for contact details

CHAPTER TEN

Fruit and Vegetables

Is a tomato just a tomato? An orange just an orange? Can mango trees produce fruit all year round and can mandarins be grown without seeds?

In the Australian fruit and vegetable industry, all things are becoming possible.

Meeting the demands of global markets means producing fruit and vegetables that are rich in flavour, vivid in colour, have longer shelf lives and have enhanced nutritional value. Collaboration between all levels of industry, from policy makers to scientists and growers, is ensuring a cohesive approach to providing international markets with world-quality produce as fresh as the day it was picked.

Horticulture in Australia is a $4.4 billion industry, with 25 per cent of all produce being exported. Australia is positioning itself to become a dominant force in world markets, working to capitalise on its sound horticultural infrastructure, proximity to significant Asian markets, abundant natural resources and technological capital. The total value of exports of fresh fruit, vegetables and nuts in 1997–98 was just under $600 million, up three per cent on the previous year, with exports of fresh fruit to Hong Kong up by 81 per cent.

The key to ongoing success is in maintaining high levels of production while keeping costs down and being ready to respond to changing consumer demands. Benchmark studies in the fruit industry have shown that in terms of labour productivity, Australia is a low-cost, efficient producer. Australia is committed to environmentally sustainable solutions to crop production, ensuring that the future is long and prosperous.

From peas and beans to broccoli and potatoes, paw paws and strawberries to apples and pears, the Australian fruit and vegetable industry is firstly characterised by its diversity. Australians have almost taken for granted the availability of crisp, fresh vegetables and ripe, juicy fruits across all seasons. Now, the benefits are being shipped overseas to countries eager for quality produce grown in clean, environmentally friendly surroundings.

Small, family-run farms are the backbone of much of the Australian fruit and vegetable industry, with many finding significant demand for their high-quality

Above: *The humble pea, or* pisum sativum, *is a long-established favourite in many cuisines throughout the world, and one of a wide variety of vegetables harvested in Australia. The part of the plant that is commonly eaten is the round, highly nutritious seed* produce in overseas markets. On the outskirts of Melbourne, Keilor Valley Gardens is a third-generation market garden now providing fresh pre-cut vegetables and salads to airline caterers in Beijing and Shanghai. With four properties covering 25 hectares, the company grows, picks, and processes the vegetables for overseas, interstate and local markets. The paddock-to-plate service is being greeted with great enthusiasm by the food service and catering industries. The added value is part of a changing philosophy in Australian fresh produce, to move from simple commodities to products with immediate appeal for domestic and global markets.

At the other end of the scale, multinational corporations such as Simplot are using Australia as a base for new developments in fruit and vegetable products. Australia's Institute of Horticultural Development (IHD) and Simplot Australia have joined forces to take advantage of emerging markets for minimally processed or fresh-cut vegetable products. Following trends in Europe and the United States, IHD and Simplot are developing fresh-cut salad and stir-fry products for Australian markets. Ingredients include vegetable medleys with broccoli and cauliflower florets that can be combined with mixes such as a cheese sauce. The venture has led to a new range of vegetable mixes under the Edgell Fresh brand.

Opposite: *Cabbages growing in a market garden at Forth, in Tasmania's fertile north-western agricultural region. Small, family-run farms are the backbone of much of the Australian fruit and vegetable industry, with many finding significant demand for their high-quality produce in overseas markets*

Above: *Pontiac potatoes freshly dug out of the ground. Potatoes are a staple food in most Western countries, and increasingly popular in the East, due to the rapid globalisation of the agri-food industry. Scientists at the Commonwealth Scientific and Industrial Research Organisation have developed a low-browning potato that is suitable for manufacturing french fries*

Around the world, there is growing demand for the freshness Australians have come to expect. Australia exports fresh oranges, lemons, limes, grapefruit and mandarins, as well as their juices, to markets mostly throughout Asia and the United States. In 1996–97, Australia exported almost 132,000 tonnes of citrus fruit. Despite volatile prices on world markets, this represented a fourfold increase on 1988–89 levels, with the growth largely due to greater demand for navel oranges and mandarins. The outlook is for exports to the United States to grow significantly, earning the industry $85 million.

Sunraysia citrus growers are looking to forge links with Japan, making a move into traditionally South African markets. After a recent tour of Japan, the Sunraysia–Osaka Economic Task Force Group was established with a view to building stronger trade relations.

Australia's $170 million fresh stone-fruit industry is also picking up its focus on exports, expanding on existing markets in Hong Kong, Singapore, Taiwan and Malaysia. The industry exports just over 10 per cent of its total production, amounting to $18 million per year.

Australia's apple industry is poised to increase the promotion of one of its best home-bred varieties, the Pink Lady. Already well received in the United Kingdom and Europe as a high-quality variety, there is considerable potential for export growth. Other popular apple varieties grown in Australia include Red Delicious and Granny Smith. The local apple industry has had the advantage of being free of the debilitating disease, Fire Blight, that has affected apples grown in several other countries around the world.

Australia's strongest fruit and vegetable export markets are in pears, citrus fruits, grapes and asparagus, but new crops are constantly being developed to meet the ever-changing tastes of consumers. Niche markets are creating invaluable opportunities for innovative growers, who are investigating the wide range of gourmet possibilities among the world's cultures.

In the Northern Territory and the Kimberley region of Western Australia, the Commonwealth Scientific and Industrial Research Organisation (CSIRO) is developing a method to induce flowering in mango plants, in order to produce fruits in otherwise hostile environments.

The CSIRO's Plant Industry Division is working on breeding melons with increased sweetness and improved storage properties for export markets. It is also developing Australia's native peach, the quandong, for commercial cultivation.

The Australian Native Bushfood Industry Council is seeking to let the world know about the extraordinary array of fruits and vegetables that have their origins in Australian soil. The British supermarket chain, Sainsburys, has confirmed that there is international demand for Australian bush foods, after making a large order at a recent tasting organised for buyers. British consumers will now have a range of new fine dining opportunities from Victorian-based Robin Bush Foods in 200 stores across the country. Produce such as appleberries, mountain pepper berries, Warrigal greens and bush tomatoes are receiving increasing orders.

From introducing the world to indigenous fruits and vegetables to catering for existing international tastes, Australian producers are listening to the desires of markets and responding with consistently high-quality produce.

Asian-vegetable growers are taking advantage of expanding local and international markets, with high demand for Chinese cabbage, Chinese broccoli, Japanese pumpkin, bok choy, Japanese ginger and several varieties of fungi. With more and more Australians attracted to the health benefits and spicy flavours of Asian cooking, the $60 million Asian-vegetable industry certainly has strong growth potential. Western Australian growers, in particular, have responded to the strong demand for Chinese cabbage. Further new markets are opening up for Australian burdock and daikon in Japan and Taiwan.

Today's well-educated consumer is eager for the healthiest products available, with a minimum of chemical use and a commitment to cleaner production techniques.

A delegation from the Japanese Ministry of International Trade to Australia found that The Original Juice Company was saving hundreds of thousands of dollars in trade-waste costs by establishing cleaner production initiatives. The company has won an Environment Protection Authority award for its world's-best-practice techniques.

Opposite (clockwise from top left): *Luscious red Jonathan apples ready for harvest; pears are another popular Australian fruit, regularly eaten fresh but just as delicious canned or made into fruit juice; nectarines, a more exotic form of the common peach, ripening into a rich red hue on the tree; wild blackberries feature highly in many modern gourmet desserts – valued for their aesthetic appeal as well as their succulence and delicious flavour*

Above: *Australian apples – not forbidden fruit, but good enough to tempt even the most discerning tastebuds!*

Overleaf: *An agricultural landscape in Tasmania. Also known variously as the 'Island State' and the 'Apple Isle', Tasmania enjoys an international reputation for its agricultural produce which is disproportionate to its small size*

Feedback from Australia's trading partners indicates that the National Standard for Organic and Bio-dynamic Produce is regarded as the world's best. All Australian organic produce must meet the Standard before it can be exported, guaranteeing minimum requirements for production, processing and labelling. It requires exporters, producers and processors of organic foods to have their operations certified and has enhanced the credibility of the organic farming industry. The Standard is used to assess organisations seeking accreditation with the Australian Quarantine and Inspection Service (AQIS). The AQIS accreditation program helped Australia to become the first country to be recognised by the European Union as an approved supplier of organic produce.

The Supermarket to Asia (STA) program is the cornerstone of the Australian Government's strategy to take advantage of growing export markets in Asia. Working closely with industry, STA's role is to improve the competitiveness of Australia's fresh and processed food exports, to take advantage of the $US16 billion per year growth in Asian food consumption. STA tackles issues across the food supply chain that are beyond the scope of any single sector, ensuring that programs are in place for a cohesive movement from seed production to on-time delivery of world-class products.

The Australian Horticultural Corporation's 'Australia Fresh' scheme provides promotional support, building the profile of the Australian fruit and vegetable industry in export markets. The scheme has the support of 31 of Australia's leading exporters who, as a group, are responsible for more than 70 per cent of Australia's fruit and vegetable exports. Australia Fresh provides a united front for exporters, giving them added strength to compete in global markets. All participants must meet the highest possible quality standards. Malaysia, Indonesia and Singapore, in particular, are already enjoying the variety, cleanliness and good tastes of Australia's fresh produce, with Australia Fresh spreading the word into more Asian markets.

Opposite and above: *An assortment of Australia's bush fruits and nuts, which have taken the international gourmet foods market by storm. Pioneered by companies such as Bush Tucker Supply Australia, the bush foods movement has benefited from the increased international demand for organic foods and new flavours*

As we approach a future with greater access to scientific knowledge, stronger cohesion between all levels of industry and a more thorough understanding of market demands, staying competitive will mean delivering products that in the past simply did not exist. We know that Asian consumers want sweeter fruits, United States consumers are seeking seedless, easy-to-peel citrus fruits, and overseas markets around the globe want pest-free produce without chemical residue. Australian scientists are responding to these demands by using a combination of genetic engineering and conventional breeding. Altering the genetic make-up of a fruit or vegetable plant can enhance its colour, flavour and firmness, or increase its resistance to pests and diseases.

Scientists at the CSIRO are tackling the problem of unsightly browning that occurs on potatoes when they are bruised during harvest, transport or storage. Although the browning is not harmful, consumers are more likely to reject affected

Bunch
$1.80
OR 90¢ BUNCH HALF
$4.50
BUNCH
OR $1.50

potatoes. By isolating the gene that causes the discolouration, scientists have been able to produce a low-browning potato intended for manufacturing french fries. The same technology is being investigated for the improvement of bananas, lettuce, apples, pears and pineapples.

Scientists are also turning their attention to the health benefits of fruit and vegetables, examining ways of enhancing or extracting medicinal properties for the prevention and treatment of diseases

Also at the CSIRO, scientists are responding to consumer demands for firm, bright-red tomatoes. The problem has been that very firm fruit often has poor flavour, so, by adding an extra copy of the gene that softens the fruit, there has been great success in developing the right balance of sugars and acids.

IHD and the Horticultural Research and Development Corporation are developing new, high-quality strawberry varieties which are adapted to Australian growing conditions. The research is aimed at producing pest- and disease-resistant fruits with improved flavours, longer shelf lives and resistance to rain damage. IHD is also working on a raspberry-breeding program, designed to extend the availability of fresh raspberries to nine months of the year. Scientists are working to produce varieties which are better adapted to Australia's mild winters, hot summers and long autumns.

A priority in all of these breeding programs is the development of fruits and vegetables that will require less chemical intervention. Significant advances in gene technology have led to improved pest resistance. New genes that inhibit the growth of apple moths have been added to certain apple varieties, reducing the need for chemical spraying.

Scientists are also turning their attention to the health benefits of fruit and vegetables, examining ways of enhancing or extracting medicinal properties for the prevention and treatment of diseases. Worldwide, the new buzzword in horticultural and pharmaceutical circles is 'nutraceuticals'. The possibilities are being investigated in research institutes around the world, with Australia capitalising on the strong reputation of garlic as a therapeutic food. An Australian company producing an

Opposite (clockwise from top left): *A vegetable seller proudly displays his selection of fresh produce, including carrots and red capsicums; excellent information and distribution links between producers and retailers mean that supermarket shelves are stacked with a huge variety of quality produce to meet consumer demand; Asian-vegetable growers are taking advantage of expanding local and international markets, with high demand for Chinese cabbage, Chinese broccoli, Japanese pumpkin, bok choy, Japanese ginger and several varieties of fungi; Chinese snow peas ready for harvesting – the pods are picked at an immature stage and eaten whole*

Above: A herb farm in Taggerty, in the Goulburn Murray Waters region in Victoria. Apart from their beauty and wonderful flavour, herbs are sought-after for their therapeutic qualities. There is an increasing interest in 'nutraceuticals', or plants with medicinal properties. The possibilities are being investigated in research institutes around the world, with Australia capitalising on the strong reputation of garlic, in particular, as a therapeutic food

extract from garlic with the potential to lower cholesterol levels is tapping into a lucrative market. Estimates are that the Australian market for production and sale of such therapies is $119 million, with the international market valued at billions of dollars. Comparisons of Australian and Chinese garlic have shown the Australian variety to have superior qualities for extraction.

With most of Australia's fresh fruits and vegetables being enjoyed by domestic markets, industry participants at all levels are now focused on export growth. As one of the world's cleanest and greenest countries, Australia's fruit and vegetable industry is becoming increasingly customer-focused, developing new produce to meet the tastes of markets across the globe. Flavour, colour, firmness, texture, nutrition and, most of all, freshness, are the priorities of Australian producers, who are committed to providing quality to the world.

Jacqui MacDonald

Australian Pure Fruits Pty Ltd

A commitment to quality, service and competitive prices have built alliances between Australian Pure Fruits and some of the world's leading food brands.

Based in the unspoilt Adelaide Hills of South Australia, the company manufactures fruit and vegetable purees and concentrates which bring natural, wholesome flavour to household name brands and leading sellers.

The expanding range includes apple, pear, strawberry, cherry, stone fruit, pumpkin and onion concentrates, processed at a plant with government quality assurance accreditation.

Produce is sourced from premium orchard and market gardening areas boasting a lush Mediterranean climate, fertile soils and quality water.

Concentrates are packaged in 200-litre asceptic bags in drums and 10- to 20-litre asceptic bags in boxes, pallet-stacked.

The company's plant also incorporates a high-speed, state-of-the art bottling line, which can handle glass and PET containers in various sizes, for both still and carbonated drinks. The plant is available on a contract basis, with clients including a well-known Asian brand of health fruit drinks now being marketed worldwide.

Above: *Quality apples ready for harvest in the Adelaide Hills*

Please refer to page 186 for contact details

Asparagus Growers of Sunraysia Inc.

A clean environment, quality management and unusual seasonal timing for export have brought success to members of Asparagus Growers of Sunraysia Incorporated.

Sunraysia is a fertile horticultural oasis, surrounding the Victorian and New South Wales border town of Mildura. In the last ten years, the region's asparagus industry has grown into a strong group of growers that produce over 1,000 tonnes of fresh green asparagus for local and overseas markets.

Asparagus from Sunraysia is freighted daily to Australian cities and numerous overseas countries during the growing season. The region's asparagus is highly regarded in the demanding Asian markets for its freshness, shelf life and flavour.

This delicious vegetable can be harvested in Sunraysia all year round. Growers are able to designate different fields to produce at varying times of the year, with the majority of production during August to October.

The area has very modern, hygienic packing sheds with excellent facilities for rapid cooling, including hydro-coolers and coolrooms.

Above: *Fresh, green asparagus from Sunraysia is delicious at any time of year*

Please refer to page 185 for contact details

Costa's

Clockwise from top left: *Activity at the Melbourne Distribution Centre; Costa's Grain Mill receiving fresh produce direct from a farm; the initial stage of the orange prepack process; loading produce for the export market; the export manager with his quality controller inspecting produce*

In the Australian market, Costa's success as a major fruit and vegetable wholesaler and distributor has been built through commitment to customer service and by creating strong relationships with suppliers.

Costa's operates major wholesale distribution centres in Melbourne and Sydney, distributing fresh fruits and vegetables across the Eastern seaboard for the supermarket trade. The centres are large, modern, well-equipped warehouses built to specification, and operate 24 hours a day, 7 days a week, to ensure produce is always at its peak of freshness for receival by customers. Deliveries are made in fast, modern, refrigerated transports.

Costa's also specialises in supplying small, premium, prepacked fruits, vegetables and flowers to supermarkets. Also part of the group are potato packing operations in Victoria and New South Wales. The supply of Queensland bananas to Australia's southern States is another major function.

The company is dedicated to continuous quality improvement. At all levels, Costa's 600 employees are educated and motivated to put quality first, alongside the satisfaction of buyers' needs and preferences. Within the framework of the international HACCP accreditation system for food safety, Costa's is extending the formal elements of ISO 9002 with the Australian SQF 2000 system to achieve the best in food quality and safety.

An integral philosophy is that good suppliers are crucial and should be supported. Much time is spent with producers, assisting them to achieve the best possible crops, and product is sourced from regular suppliers. A program to train and accredit 400 suppliers to SQF 2000 standards is already well underway, on the basis that quality food begins on the farm.

The well-deserved success of Costa's Fresh Produce is based on customer satisfaction, the best suppliers, putting quality first, and careful attention to management of the whole supply chain.

SOUTHERN CROSS

As a leading Australian fruit and vegetable wholesaler and distributor, Costa's has capitalised on Australia's international reputation for premium agricultural produce based on good growing conditions and a clean environment.

During the 1980s, Australia's role in supplying fresh agricultural produce to off-season markets in the northern hemisphere grew, while demand from Asia also increased substantially. Responding to these opportunities, Costa's developed a subsidiary named Southern Cross Overseas Proprietary Limited, based in Geelong, Victoria. Southern Cross began exporting in 1988, and it now serves around 40 countries.

Southern Cross Grains, specialising in grains, pulses and potatoes, began operations in 1993 and complements Southern Cross Overseas, which focuses on fresh fruits and vegetables. Export markets have been built on the basis of relationships – long-term and trusting associations with customers, and partnerships with suppliers.

By concentrating on meeting the needs of customers, Southern Cross Grains has developed niche markets virtually around the world. It supplies highest quality horsefeed to the largest racehorse owners in the world, and to both the Hong Kong and Macau Jockey Clubs. Sprouting peas from Southern Cross are highly prized in Malaysia, while potatoes find ready markets in South Korea and Hong Kong. Seeds are exported to Japan, forage seeds to Libya, and specialist seeds to the Middle East, together with feed supplements and minerals.

Southern Cross Overseas views the whole of Asia as its home market and the source of future growth, with air or sea shipments of fruits and vegetables departing every day for any of the Asian countries. Export fruits are of high value, including stone fruits, citrus fruits, grapes, mangoes and melons, while vegetable exports range across onions, potatoes, broccoli and cauliflowers to Chinese cabbage and many other lines. Suppliers and market buyers are chosen for their intimate knowledge of the Asian market.

With their emphasis on customer satisfaction and best-quality produce, Costa's and its subsidiary, Southern Cross, are well positioned to lead Australia's fruit and vegetable industry into the new millennium.

Please refer to page 190 for contact details

ORANGE JUICE SQUEEZED DAILY
The Outback Juice Company Pty Ltd

This page and opposite: *At every level in the food supply chain, Moraitis is committed to being the 'Leader in Fresh Produce'*

Moraitis

LEADERS IN FRESH PRODUCE

Nicholas Moraitis has always had a clear objective in business – simply to provide his customers with the very best in quality produce, as well as an uncompromising level of service. It was with this objective and business philosophy that he built the company today known as Moraitis Fresh, which has become one of the country's leaders in the fresh produce industry.

Moraitis Fresh has grown over the years, with operations spanning throughout the east coast of Australia, covering wholesaling, packaging, processing and growing of fresh produce. The company has always focused on food safety and quality, with links along the supply chain providing valuable feedback from all major retailers to growers.

The company has its own full-time marketing department, which is constantly developing strategic campaigns aimed at broadening consumers' perspectives of potatoes and associated products, promoting the industry in the face of changing consumer habits.

SCOPE OF MORAITIS' OPERATIONS

As one of the largest produce wholesalers in Australia, Moraitis Fresh has the capacity to supply some 4,000 tonnes of fresh produce each week throughout Australia.

Apart from the traditional company base of wholesale hard produce lines, Moraitis Fresh has expanded its range to include fresh fruit and vegetables, citrus juices, fresh, processed, safe vegetables and salads, and a variety of prepackaged lines.

Today, the company is more focused than ever, driven by the principles on which it was founded. The unique leadership team comprises Nicholas Moraitis as Chairman and Chief Executive Officer, along with sons Stephen and Paul as company directors – each of whom plays an integral hands-on role in the company's every-day operations and overall success.

EXAMPLES OF MORAITIS' INITIATIVES

Growers are invited to commit to the 'Moraitis Preferred Supplier Program', which outlines the standards of quality management required on-farm to ensure that wholesome, safe produce reaches Moraitis sites according to consumer specifications for size and quality, with minimum defects.

Moraitis Agricultural is the company's very own farm, 'Belvedere', a new undertaking with up-to-date facilities and equipment, as well as management practices which will undoubtedly establish it as a leading quality grower in the hard produce area.

Moraitis Fresh (Riverina) serves as a direct link with growers in this most important district, sourcing and packing the area's premium product, whilst offering assistance and advice to growers.

FreshPak Australia, a joint venture between the Atherton Potato Growers Co-operative and Moraitis Fresh, is located in Far North Queensland, and draws potatoes from one of Australia's premier growing areas, equipped with its own on-site grading and packing facilities.

QUALITY SYSTEMS

Quality systems are implemented throughout the business, and external audits are conducted by the internationally renowned auditors, SGS Limited. The focus on food safety and quality is obviously very high at the value-adding plant in Sydney, where a range of salads and stir-fries are packed in controlled atmosphere packaging as ready-to-eat fresh produce.

State-of-the-art technology is continually assessed, both here and overseas, to identify the most efficient processing, grading and packaging equipment that will protect produce safety and quality. As Nicholas Moraitis states: 'Our commitment is demonstrated by our expenditure of over $20,000 a week on food safety and quality at the value-adding plant alone – we are in this for the long haul'.

This commitment is what sets apart the Moraitis group of companies.

The wholesale operations are equipped to supply Australia's major retailers and independent outlets with their specific requirements for a range of prepacked bags, trays and punnets of fruit and vegetables. Large volumes of potatoes and onions are sorted and packed 'around the clock' for timely delivery by the Moraitis transport fleet.

Market stands in Brisbane, Sydney and Melbourne provide direct contact with agents and resellers of fruit and vegetable lines, as well as the opportunity to wholesale large volumes of fresh produce.

Product identification and traceability has probably been the biggest challenge facing Moraitis and the industry as a whole. By integrating the supply chain, Moraitis is better equipped than most to monitor the quality of fresh produce.

There are no short cuts to implementing a food safety and quality management system. With a business the size of Moraitis, it requires a consistent systematic approach focused on continuous improvement.

The Moraitis family have the commitment and capability to achieve their version of 'Leaders in Fresh Produce' well into the 21st century.

Please refer to page 195 for contact details

Ardmona Fresh Fruit Packers Pty Ltd

'THE VARAPODIO FAMILY'

In 1926 Rocco Varapodio migrated from Italy and settled in the Goulburn Valley, working for some years before establishing his own market garden in 1937. After developing the fruit-growing property, Rocco and his son Santo began packing their own fruit by 1950, thereby starting the tradition of 'Varapodio' quality.

Today, Santo and his son, Rocky, operate 250 hectares of orchard, producing approximately 10,000 tonnes of apples, pears, apricots, plums and white- and yellow-fleshed peaches and nectarines.

Innovation and technology are integral parts of the Varapodios' business strategy. From planting to marketing, the fruit is nurtured by highly trained staff. Horticulturists and entomologists monitor the following areas:

- soil make-up;
- planting of the trees;
- tree nutrient requirements;
- crop quality and variety;
- limiting pesticide use through the use of natural predators; and
- high pruning standards.

The 'tatura trellis' method of training trees along wires produces earlier, better yields of higher-quality fruit. Computerised storage and grading of fruit ensures excellent outturns of high-quality fruit to the consumer.

Cold storage is an essential part of maintaining quality. The fruit is cooled to zero degrees in one of 20 coolrooms immediately after harvest. The atmosphere is controlled to reduce oxygen and carbon dioxide, slowing the respiration of the fruit and retarding ripening. Apples and pears can be stored in a controlled atmosphere for up to ten months without losing quality.

Fruit quality is monitored by a quality assurance manager from harvest until it leaves the packhouse. In 1993, Ardmona Fresh Fruit Packers was the first orchard in the Goulburn Valley to receive the Australian Quarantine and Inspection Service's certification assurance.

Varapodio's quality fruit is distributed to all Australian markets. Since commencing exports in the late 1970s, the Varapodios have built up considerable experience of serving international customers in Europe, the United States and South East Asia, and are constantly looking to expand their overseas markets.

The Varapodios produce great Australian fruit, but as they say: 'The proof is in the tasting!'

Clockwise from left: *peach blossoms flourishing in the lush Goulburn Valley; Australian Packham pears, world-renowned for their excellent flavour; crisp Granny Smith apples; quality white-fleshed nectarines*

Please refer to page 184 for contact details

edp® Packaging and Handling Systems

In 1947, Eric Douglas Parsons designed and built a fruit-picking ladder at the request of a local orchardist at Mooroopna, in the heart of Australia's Goulburn Valley fruit-growing belt. Today, the company that bears his initials has become an international supplier of packaging and handling systems for fresh fruit and vegetables.

edp® manufactures a range of fruit and vegetable packaging and grading machinery which suits Australian conditions, fruit and vegetable varieties and packaging requirements at the retail level. The company offers new developments in packaging with extensive international travel undertaken by many of its staff.

edp® is involved in the global economy, with the company exporting equipment to every continent and being equally competitive with other overseas manufacturers, both on the domestic market as well as internationally.

There is a heavy emphasis at edp® on raising clients' productivity by including features that lead to faster grading with less damage, by keeping product travel to a minimum. For example, sizer designs maintain high accuracy while minimising product damage, goose-neck delivery on elevators ensures that product rolls off rather than being thrown, and dual-bin fillers have a unique tilt design, allowing product to roll gently into bins which automatically lower as they fill.

edp® supplies complete turnkey grading systems for vegetables, including bulk receival hoppers, pre-inspection, cleaning units, soil removal, onion top and tailing systems and grading facilities, and potato dry brushing or washing. At the packaging stage, the company offers the world-renowned Lynx computerised weigher, and the Electro-bagger and Minibagger weighing and packaging machines. Between them, these systems handle conveyance and consumer packs.

edp® is about to create a 'world first' with its new Electronic Vegetable Sizing System.

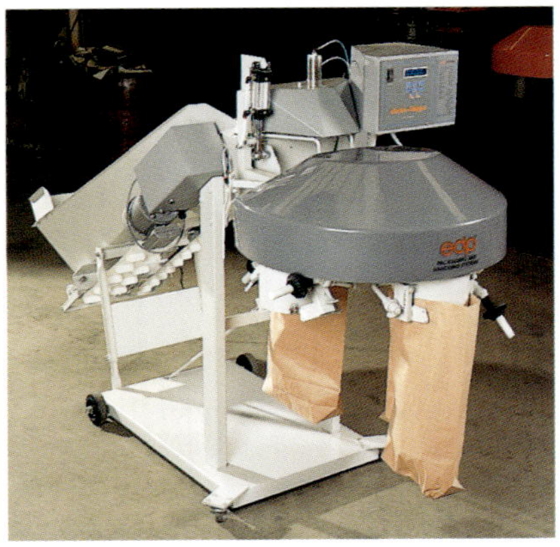

Top: *Designed to maximise productivity over smaller runs, the edp® Minibagger automatically fills bags quickly, then trim-feeds each to the required weight. It is shown here with the optional Empac GH-10 automatic net bagger – ideal for packing tomatoes, peppers, apples, Kiwi fruit and soft citrus fruit*

Centre: *The new edp® Powerwate Powerladders have been setting the pace in the fruit-picking and pruning industry for more than 17 years*

Left: *The edp® Electro-Bagger is designed for bagging potatoes, onions, lucerne cubes or produce larger than 25 millimetres in diameter*

Please refer to page 191 for contact details

Gartner Farms

From variety selection to packaging which ensures delivery of fresh, quality produce, attention to detail is the priority for Gartner Farms.

The family-owned enterprise harnesses the rich soils, pure, abundant water and mild climate of South Australia's fertile south-east to produce onions, potatoes and sweet corn with superior keeping qualities. Backed by experience and extensive trials, Gartner varies its produce output according to the prevailing market demand for specific vegetables.

Vegetables are graded and packed carefully in a fully accredited packhouse, and loaded and transported directly to meet precise customer specifications and the certification requirements of countries of destination.

Gartner has built its export business on the principle of getting to know exactly what is wanted by its customers in Asia and further afield. The family-owned company pays close attention to consistency, quality and long-term reliability, and actively seeks market feedback as part of its commitment to customer service.

Above: *David and Michael Gartner checking out newly planted onions in the fertile soils of Glenroy*
Right: *A bumper crop of canola being inspected by Greg Gartner with overseer, Toby the dog*

Please refer to page 192 for contact details

Melbourne Markets

The National Flower Centre, located at Melbourne Markets, provides modern facilities for the distribution of Victoria's cut-flower industry

Melbourne Markets is Victoria's largest wholesale fresh produce complex, with an annual turnover of $1.2 billion. Incorporating the Wholesale Fruit and Vegetable Market and the National Flower Centre of Australia, it is just five minutes from the city of Melbourne, and covers 30.1 hectares.

With up to 600 growers' stands and 140 wholesalers under one roof, the fruit and vegetable market is one of the biggest markets in the world. Produce arrives fresh from growers' farms, often only hours since it was picked. Located close to some of the most fertile growing areas in Australia, the market is able to offer an outstanding level of variety, freshness and quality.

The flower centre is the first fully temperature-controlled market in Australia, and is pollution-free. It caters for the substantial Victorian flower industry, which produces about 42 per cent of the $350 million-per-year national flower market.

Both the fruit and vegetable and flower markets have adopted quality assurance programs to ensure high quality in supply and exports.

Please refer to page 201 for contact details

Harvest Moon

The rich, red volcanic soils of Tasmania's north-west coast are the basis of a multi-million-dollar vegetable growing and processing industry that has earned Tasmania the reputation as Australia's 'vegetable basket'.

Based in the fertile Forth Valley near Devonport, Forth Farm Produce, which produces the successful Harvest Moon brand, is at the leading edge of quality fresh vegetable production for local and interstate consumption and export.

Harvest Moon has achieved its enviable position by capitalising on the State's natural advantages of fertile soils, mild climate and clean water, to produce vegetables that are renowned for their flavour and keeping qualities.

While the State's jealously guarded reputation as being clean and green has been a valuable tool in winning and maintaining overseas markets, the company has backed that up by following a rigorous quality management philosophy throughout the supply chain to ensure that customer expectations are fulfilled.

In conjunction with its customers, Harvest Moon has developed detailed product specifications for all its product lines. Specifications are tailored to suit the requirements of individual customers. While the company owns or leases some land for production, the bulk of its vegetables are grown by contract growers, who are rewarded for high-quality production.

Major product lines for supply to both interstate and overseas markets include broccoli, carrots and potatoes (South-East Asia), kabocha (Japan), onions (Europe and Asia) and swedes (United Kingdom). Smaller quantities of products, including parsnips, celery, sweet corn, beans, leeks, brussels sprouts, cabbage, cauliflower and zucchini, are grown for the Tasmanian market.

Each of the major crop lines has a product manager, whose role is to ensure that systems are in place to enable customers' and producers' expectations to be met.

Sales staff have specific responsibility for particular crops and regions, allowing staff members to develop close working relationships with customers and familiarity with the vegetable lines they are selling. Field staff take responsibility for grower selection, testing and contract ground preparation and harvesting.

Top to bottom: *Fertile soils, mild climate and clean water provide the perfect growing environment for Harvest Moon's fresh produce; Tasmania's north-west coast – the 'vegetable basket' of Australia; Harvest Moon supplies local, interstate and overseas markets*

With an eye on consistency of production and an ear to customer requirements, Forth Farm Produce is working to ensure the Harvest Moon brand can be found on supermarket shelves around the world, throughout the year.

Please refer to page 193 for contact details

Perfecta Produce

With fewer than half a million people and less than one per cent of Australia's land area, Tasmania – with its natural advantage of isolation as a defence against pest and disease – has an international reputation that far exceeds its size.

Located on the State's north-west coast, Perfecta Produce is a major exporter of onions, swedes, carrots and shallots to worldwide destinations including Japan, South-East Asia and Europe.

Established in 1984, the company has built a reputation for customer satisfaction based on a history of quality products, quick response and timely service.

Stringent quality controls, committed personnel, and the pride and passion to specialise in each product have set the Perfecta brand as the standard of excellence with customers worldwide.

The company is Tasmania's longest-established onion producer and exporter, with its most exciting development in recent years being the growing and exporting of fresh cherries.

Successive plantings have resulted in a total of more than 20,000 cherry trees, covering nine hectares of high-density orchard. The orchard area is covered by netting for protection against birds, and extensive rain cover protects the crop during the cherry harvest season, guaranteeing supply.

With an expected yield of some 200 tonnes per year, Perfecta leads the State in the production of this highly valued crop.

Rather than growing a product and then seeking a market, Perfecta determines what markets require and then produces according to those needs. This is a philosophy which has seen the company help the island State on its way to becoming an emblem for the finest foods in the furthest destinations – and is part of the Perfecta trademark.

Top: *An aerial view over Perfecta Produce and the rich volcanic soils of north-west Tasmania's growing region*

Centre: *Tasmanian Creamgold onions packed awaiting export to Europe*

Left: *Fresh 'Bing' cherries ready for picking*

Please refer to page 198 for contact details

Woolworths Limited

When Australians do their household shopping, chances are it will be at Woolworths. The company is the largest food retailer in Australia and also a major retailer of general merchandise, with 1,186 retail outlets nationally. Around 525 Woolworths supermarkets nationwide supply 34 per cent of the country's total grocery market.

Known as 'The Fresh Food People', Woolworths is committed to providing the highest quality and broadest range of competitively priced fresh food and groceries, as the buying agent for over nine million customers around Australia each week.

The freshness of produce is guaranteed by Woolworths' involvement 'from paddock to plate'. Suppliers and manufacturers must meet the company's strict quality specifications during the production and delivery of food, ensuring customers can absolutely depend on the safety of the food they buy.

To maintain its predominance, Woolworths is constantly instituting innovative new systems and technologies. It is a major player in the burgeoning meal solutions market, now providing over 400 prepared and semi-prepared lines. Sophisticated databases enable Woolworths to ensure that the ranges carried in each store meet the particular needs of the local community.

Woolworths was the first Australian supermarket company to institute Internet shopping, establishing the facility initially at its Eastwood, Sydney store in March 1998 and rapidly expanding it to surrounding areas. In 1999, the aim is to have a Woolworths 'Homeshop' presence in all Australian States and Territories.

The implementation of innovative techniques to extend the quality and freshness of produce is also a Woolworths objective. Already on the supermarket shelves is a 'perfect' vine-ripened tomato developed using Israeli-adapted technology.

Now, international customers are benefiting from Woolworths' experience as Australia's largest retailer of fruit, vegetables and meats. In recent years, Woolworths' export division has been sending superior quality Australian produce around the globe. Woolworths is the largest exporter of Australian pork and also exports fruit and vegetables to many overseas countries.

Top: *Woolworths 'Metro' store at Newtown, New South Wales, specifically designed for the urban customer*

Middle and bottom: *Customers in some Sydney suburbs now have the option of Internet-based shopping at Woolworths' new website: www.woolworths.com.au*

Please refer to page 205 for contact details

CHAPTER ELEVEN

Wines and Beverages

WINES FOR ALL SEASONS

The Australian wine industry is a remarkable international success story.

Australian wines are regularly winning awards all over the world, beating nations which have thousands of years of winemaking tradition.

Not surprisingly, the industry is enjoying an unprecedented period of growth – investment in vineyards, wineries and associated marketing has reached record levels, and continuing growth potential appears to be assured. This is still a relatively untapped export market, and conservative forecasts indicate that exports may triple in less than a decade.

A unique blueprint, titled *Strategy 2025 – The Australian Wine Industry*, was launched in mid-1996 by industry leaders. This ambitious vision statement maps, in some detail, a plan for shifting the industry to a new level of production.

Under the plan's guidelines, the industry is to establish itself firmly as one of the world's major wine producers. Exports are forecast to rise from $450 million per year currently (representing just over one-quarter of total production) to about $2.5 billion in 2025 (representing over half of total production). In percentage terms, this means that Australia's share of the total value of the international wine trade is projected to rise from about 3.5 per cent currently to about 6.5 per cent in 2025.

Behind the forecasts are a series of strategies, which include $5 billion in new investment for the industry. New plantings are planned at 40,000 hectares, double the winegrowing area in 1996, and wine-processing capacity is similarly forecast to double the present levels of about 600 million litres per year.

The result, according to the 2025 vision statement, will be to make Australia 'the most influential and profitable supplier of branded wines, pioneering wine as a universal first-choice lifestyle beverage'.

Above: *A premium-quality Australian wine adds the perfect finishing touch to a gourmet meal prepared with fresh, natural Australian produce. Australian wines are regularly winning awards all over the world, beating nations with thousands of years of winemaking tradition*

Overleaf: *Grape-pickers at work in a vineyard in the Barossa Valley, South Australia. The Barossa is widely regarded as Australia's premier wine-producing region, home to well-known wineries such as Orlando Wyndham and Chateau Yaldara*

'The industry's future lies primarily in branded wine products which reflect the distinctiveness of variety, region and producer.'

At the time of the launch, supporters of the 2025 vision conceded that revealing such a plan had never been done in Australia, but maintained that the effort would be worth the risk. To prove them right, most of the forecasts are being realised ahead of schedule just a few years into the plan.

It had been estimated that to reach the 2025 goals, 1,500 hectares of wine grapes needed to be planted each year – but in the two years since the statement was released, estimated new plantings have reached a massive 14,000 hectares. At this rate, it also appears that projections that the annual vintage would reach one million tonnes by 2001 will be fulfilled two years ahead of time.

Now that the 2025 vision is clearly achievable, the new challenge is to ensure that the pace of growth does not exceed sustainable levels. To this end, viticultural consultants around the country are keeping a close eye on transactional trends and practices within the industry.

Rosemount Estates, one of the larger winemaking companies in Australia, is committed to building new vineyards, rather than expanding contracts with outside vineyards. Marketing Director Neal Hadley is confident that the company will be able to meet the goals of the 2025 vision. He says, 'We can and will match our marketing targets to our production, we are certain of that – after all, if we don't we'll bankrupt ourselves'. Rosemount has tagged Asia (in particular, Japan), the United States and Europe (in particular, Germany), as markets with the best growth potential for its wines.

Top: *Wine is stored in barrels with an ideal capacity of 225 litres. These barrels, made of oak, are ideal for fermentation and maturation and for the evolution of complex flavours and textures. New oak is a vital component of high-quality, modern winemaking*

Above: *A chardonnay budburst in spring. The vine has been pruned to produce precisely controlled yields. Each budburst will become a new cane, and from each cane will grow two to three bunches of grapes. A vine's capacity to produce grapes comes from the latent energy within each bud – only new growth produces grapes*

Opposite: *Grapes being unloaded into a crusher at Penfolds Winery in South Australia's Barossa Valley*

Japan, says Mr Hadley, is a particularly interesting market. '(The Japanese) are a new but affluent market for wine, and have been purchasing the top-growth wines from France and the like.' However, he adds, they are becoming more value-conscious and are now 'turning to the big-name Australian red wines, which can deliver similar exceptional quality, at a far cheaper price'.

'We believe the market potential here is enormous.'

The United States and Germany are no less interesting to Australian wine companies – the United States due to its sheer size and a wealth of wine-drinkers with developed tastes, and Germany because of its population of affluent consumers.

The United States, as the world's third-largest wine import market, is currently worth a little over $50 million per year to Australian wine exporters – so there is clear potential for much greater growth.

Germany, which imports a little over $15 million worth of Australian wines each year, is the world's largest wine importer by volume. Its obvious potential as a growth market is enhanced by the fact that Australian wines enjoy an extremely positive image in that country.

5
2
PWL
PWL

Japan, with a very small local wine industry, is a largely untapped importer, and is worth a fraction of the American and German markets at the moment.

However, the backbone of Australia's wine exports has traditionally been the United Kingdom – another key market profiled in the 2025 vision document. Australia's wine exports to the United Kingdom are currently worth almost $200 million per year. Australia is perfectly positioned to target the growth areas in this market, which are predominantly for medium-priced wines.

Under the 2025 vision statement, the United Kingdom will be overtaken eventually by the United States as the major export destination for Australian wines.

Although the importance of overseas markets is undeniable, the domestic wine market is not to be forgotten as a major growth area for Australian wines. In terms of per capita consumption, the local market is forecast to grow from the present levels of just over 18 litres to 22 litres per year, with an expected annual worth of about $2 billion by 2025.

Having set their goals, industry leaders now have their heads down to harness the growth that is already underway.

Australia's unique climate is ideal for growing wine in many different regions. In addition, it is world-renowned for innovating technology in growing grapes and making wine.

As we head towards the new millennium, the biggest challenge remaining for Australia's burgeoning wine industry may be to perfect its skills in marketing the final, bottled product.

THE BEVERAGES INDUSTRY

Australia's beverage industry includes a multitude of fruit juices, in particular those processed from citrus fruit.

Australia's citrus industry has faced challenging world market conditions in the last few years. However, industry leaders are optimistic that these conditions have peaked, and most are now looking forward to a period of consolidation.

In recent years, the industry has implemented significant improvements in its processing and production efficiency. These improvements are now expected to lead it into a new period of sustained growth. New markets are opening for Australian-made fruit juices all over the world and prices are on the way up.

Indeed, many observers contend that Australia can now outperform the world's best exporters in critical markets such as South-East Asia and the Middle East. Australian juice processors and producers are finding innovative ways of beating cheaper foreign juice products by producing top-quality products aimed at the more discerning consumer.

A paper compiled by the Australian Bureau of Agricultural and Resource Economics, released in mid-1998, highlights the challenges that have been faced by the local citrus industry, as cheap imported orange juice grew from under one-third of domestic sales in 1989–90 to over two-thirds eight years later.

More than 600,000 valencia trees (which largely produce fruit for the juice market) were removed in 1994–95 and 1996–97 as growers in irrigation areas opted instead for fresh fruit trees, such as navel oranges and mandarins, or alternative crops, such as wine grapes.

The shift was largely due to a glut of juice on the world market, due to overproduction, particularly in Brazil. This caused a significant fall in the prices commanded by valencia oranges around the world.

The Australian citrus juice industry's response to these challenging conditions has been to focus its collective mind on where its strengths lie. In a signal for the future of the industry, the ten years leading to 1997 saw a quadrupling in the production of non-juice fresh citrus fruits (particularly whole oranges and mandarins), mostly bound for Malaysia, Singapore, Hong Kong, Japan, Indonesia and even the United States, in the latter's off-season. Exports of fresh citrus fruits topped $130 million in the 1996–97 financial year.

Today, Australia is still a relatively small producer on the world citrus beverage market. In terms of both processing citrus and fresh citrus, Australia accounted for just one per cent of total world production.

While much of the Asian growth, in particular, has slowed in the latter half of the 1990s, due to a regional economic downturn, exporters remain confident that a

Left: *Some wines are kept in temperature-controlled, stainless steel tanks, which have replaced wooden ones. The temperature is warmer at the top of the tank than at the bottom; to ensure an even temperature, the tanks are tumbled. The wine is stored in these tanks for less than a year*

Above: *A worker hand-harvesting grapes, which are suitable to be eaten fresh, made into wine or processed as grape juice – one of the most popular fruit juices. An alternative method of harvesting is to use machinery. Machine-harvesting can be done at night, maintaining the aromatic freshness of the fruit*

Overleaf: *The citrus juice industry's response to challenging world market conditions has been to focus its collective mind on where its strengths lie. While the regional economic downturn has slowed export growth to Asia in the latter half of the 1990s, exporters remain confident that a critical start has been made in getting Australian produce established in the various regional markets*

The Fresh

critical start has been made in getting Australian produce established in the various markets, particularly given the continuation of the medium-term trend in the decline of the Australian dollar.

The author of a report by the Australian Bureau of Agricultural Resource Economics, Ms Cathryn Rees, states that there is a renewed future in the citrus industry, thanks to the removal of government regulation.

Growers continue to work with government bodies by identifying appropriate measures which the government bodies could take to assist the industry, and desirable levels of government intervention. Areas upon which the industry has focused particular attention are origin labelling laws and sales taxes on products. Some growers assert that any lowering of tariffs on juice imports must be accompanied by allowances to encourage local growers to stay in the industry.

In the middle of the industry are fruit juice processors, who are communicating a desire to take a more co-operative approach.

One of the largest fruit juice processors is Berri Limited, which processes one in five oranges in Australia and operates nine processing plants from Darwin to Melbourne. Berri's Manager of Corporate and Legal Affairs, Ashley Roff, says that processors have been extraordinarily successful in expanding consumption in the past decade, and that there is no end in sight to this trend.

'We are learning to be more competitive through innovation of product – such as "no-added-sugar juice", and through increased efficiencies', Mr Roff states.

'The growth in imported product is, to some extent, a reflection of the expanded market we have been able to engineer.

'We are getting better at producing juice in particular, and it would be dangerous to believe the decline in juice prices will continue forever, so I believe going into the future we are well placed at all levels of the industry.'

Top: *Oranges are a mainstay of the citrus beverage industry, with orange juice maintaining its position as a favourite fruit juice on breakfast tables around the world*
Middle: *Oranges freshly harvested at a farm in Wiluna, Western Australia*
Bottom: *An aerial view of citrus and apricot farms between Morgan and Waikerie, on the Murray River in South Australia*

Many citrus juice processors are now offering long-term, fixed-price contracts, meaning that they can source products at a steady price, suiting processors, growers and the bank managers of both parties.

Most are looking to more lucrative export markets as the key to fulfilling growth plans. To this end, Australia's reputation as a clean and environmentally friendly producer of top-quality foods and beverages is standing the industry in good stead.

Tim Stevens

Berri Limited

Healthy living and good food are part of the Australian way of life. In a country where consumers demand freshness and quality, successful companies constantly strive to exceed the expectations of their customers. One company with such a commitment is Berri Limited.

During World War II, a number of co-operatives in the Riverland district of south-eastern Australia joined together to supply fruit juice to the armed forces under the 'Berri' brand. Fifty years later, Berri markets a range of natural beverages enjoyed by millions of Australian families.

Berri is Australian-owned. It employs around 1,200 people nationally, and each year processes around 100,000 tonnes of fresh citrus fruit for juice at nine sites around Australia. From its humble beginnings, the company has grown to become Australia's number one fruit juice producer and an increasingly important manufacturer and marketer of fruit-based and other nutritional convenience foods.

In the international arena, over 30 years of significant export activity is the foundation for Berri's present trade with 20 countries, primarily in Asia and the Middle East. 'Berri', the leading imported juice brand in a number of countries, can be seen on supermarket shelves from Jakarta to Moscow.

Berri is committed to producing the freshest and healthiest fruit juice products. The company is pioneering quality management systems throughout its processing operations. Hazard Analysis Critical Control Point principles for food safety were introduced in 1998. This was a natural progression from ISO 9002 accreditation and adoption of cold chain management practice, in association with major retailer and wholesaler customers.

Berri Limited is focused on the future. It understands the importance of quality, freshness and product range. It is this ethos that drives the company and will guarantee its continued success.

Top: *Harvest time at a Riverland orchard*

Middle: *Sunset in the Riverland – the area's climate provides ideal conditions for citrus fruit growth*

Bottom: *Berri – the leading brand in the long-life juice category*

Please refer to page 187 for contact details

Bundaberg Brewed Drinks

A family company in Bundaberg, Queensland, is working overtime to meet a growing demand for some of the world's oldest and healthiest soft drinks. Bundaberg Brewed Drinks manufactures a range of naturally brewed beverages, including the popular Bundaberg Ginger Beer, which sells over 20 million bottles a year.

Bundaberg Ginger Beer is one of the few ginger beers still manufactured in the traditional way, from fresh ginger, local cane sugar, brewer's yeast and pure underground water, producing a delicious non-alcoholic drink, rich in proteins, amino acids and Vitamins B12 and B6.

The same high-quality ginger is used in many herbal medicines to improve circulation and digestion, and to relieve nausea, motion sickness, coughs, colds and rheumatism.

Served in a traditional recyclable stubby, Bundaberg Ginger Beer dominates the local ginger beer market and has become a popular export to New Zealand. It is now opening new markets in the United States, Asia and the Middle East.

Bundaberg Brewed Drinks has won the prestigious Export Award for its whole range of naturally brewed soft drinks, which also includes Bundaberg Root Beer, 2-Up Horehound and Bundaberg Lemon, Lime and Bitters.

Bundaberg Brewed Drinks is owned and managed by the Fleming family, and is now in its third generation of Fleming family ownership. The company's range of popular brewed soft drinks are still manufactured in the family-owned factory in Bundaberg.

The company invests heavily in research and development, and is a world leader in natural brewing technology.

Cliff Fleming, the company's Managing Director, articulates a clear and simple focus for the business in his shared mission:

To support our people to offer leading distributors, retailers and consumers the world's leading value, traditional style, naturally brewed, non-alcoholic beverages from the finest quality ingredients, using the best suppliers and marketing strategies.

Top to bottom: *Bundaberg Ginger Beer heads up a range of popular naturally brewed drinks, all manufactured in a family-owned factory in Bundaberg, Queensland*

Please refer to page 187 for contact details

Roberts Estate Wines

A TOAST TO EXPORT GROWTH

Australian wine is recognised around the world for its quality and consistency. Forecasts predict that export sales will reach A$1 billion per annum by the year 2000.

Investment in new vineyards and production facilities is at record levels. The Australian wine industry is a major player in an export sector which is the envy of other countries.

Five years ago, Irymple Packing Proprietary Limited (the Irymple Group) recognised the potential of the Australian wine industry. As a result, the company began to explore the possibility of further diversifying from existing business activities.

Irymple Group was established at the turn of the century. The company built a solid reputation on dried-fruit packing, citrus and vegetable juice production and the manufacture of concentrates and paste products. In the early 1990s, focus on vegetable juice concentration led to the company becoming the largest exporter of carrot and capsicum concentrate to Asia.

In 1996, the company established a substantial grape-crushing facility. Although this facility was initially developed to produce premium juice for concentration and de-ionisation, the potential to value-add through wine production soon became apparent. Installation of bulk storage facilities in 1997 took the company one step closer to fully fledged wine production.

Built from the ground up, the new winery at Merbein (15 kilometres west of Mildura) features the latest technology for use in wine production. Today, the company offers a service to other producers by supplying high-quality concentrate and de-ionised juice, and providing contract crushing and winemaking facilities.

The company is committing significant resources to the export market. Following the 1998 vintage, Roberts Estate Wines was launched. Marketing two ranges, Kombacy for the United Kingdom and Denbeigh for all other markets, the Irymple Group is taking its wines to the world.

Please refer to page 200 for contact details

Top to bottom: *Launched in 1998, Roberts Estate Wines are marketing premium quality wines under two ranges – Kombacy for the United Kingdom and Denbeigh for all other markets*

Gartner Wines

One of the newest winemaking ventures in Australia's most prestigious winegrowing region is based on a solid foundation of experience.

Gartner Wines produces premium red and white varieties from mainly cabernet sauvignon, chardonnay and merlot grapes grown in South Australia's world-famous Coonawarra region.

Primary producers for generations, the Gartner family has been supplying quality grapes to some of Australia's most respected premium labels, which are enjoyed by wine-lovers around the globe. Stretching over more than 440 hectares, the vineyards exploit the fertile soils, pure water, mild climate and world's-best viticultural practices which characterise the Coonawarra region.

The Gartners also operate the Heathcote Winery, an established winery in central Victoria, which is expanding to meet demand and is renowned for its high-quality shiraz, viognier and chardonnay wines.

Above: *Another exceptional vintage in the making! Director of Gartner Wines, Michael Gartner, inspects healthy budburst with vineyard manager, Steve Reyne, at their Coonawarra vineyard*

Please refer to page 192 for contact details

The Original Juice Company

Established in 1988, The Original Juice Company has a reputation built on quality and know-how.

The company's modern production facility is the largest purpose-built fresh fruit juice extraction plant in the southern hemisphere. This three-hectare site manufactures consistently high-quality, fresh, chilled fruit juice and long-life pure premium – none of which is reconstituted from concentrate, and has facilitated innovation in unique, exciting new beverages.

The Original Juice Company has developed an expanding national distribution vendor network, which ensures that the integrity and freshness of the product is maintained throughout the cold chain. Increasingly, the company's expansion into international markets is a result of its focus on quality, consistency and value. The co-ordinated branding, merchandising and marketing campaigns adopted create high recognition and brand recall.

The Original Juice Company is an energetic family business that has built proud and long-lasting links with growers, suppliers, distributors and customers. It has achieved segment leadership in Australia and accelerated expansion overseas.

Top: *The Original Juice Company has expanded into international markets*

Bottom: *Premium freshly-squeezed orange juice – a fresh, natural taste produced from high-quality fruit*

Please refer to page 203 for contact details

Jacob's Creek®

THE CREEK THAT FLOODED THE WORLD

When Johann Gramp planted his first vine cuttings beside Jacob's Creek in South Australia, he set in motion an international success story.

The Bavarian settler's endeavours in 1847 marked the birth of Australia's premium wine region, the Barossa Valley, and a wine which today is consumed around the world.

Jacob's Creek has become Australia's top-selling and most exported brand of bottled wine, as well as the best-selling Australian wine in numerous Asian and European countries.

Produced by Orlando Wines, this smooth, flavoursome wine has succeeded globally through a commitment to quality, excellent value and consistency.

'To create a long-term brand internationally you need the right price, packaging and product positioning – and that is what we have established', says sales and marketing director, Stephen Couche.

Assisting the company's push into world markets is a formidable international distribution network, including unique marketing alliances in Asia, and a significant expansion program to boost production.

Bottom: *Jacob's Creek, 'Australia's Top Drop', is the number-one-selling bottled wine brand in Australia – exported to more than forty countries worldwide. Its national and international reputation continues to expand as a result of its consistently high-quality and award-winning red and white wines*

Please refer to page 197 for contact details

PATE

CHAPTER TWELVE

Processed Foods

AUSTRALIA'S LARGEST MANUFACTURING SECTOR

The processed foods and beverages industry is the largest manufacturing sector in Australia and a front-runner in claiming a bigger slice of international markets for producers and exporters.

The industry has an annual turnover of about $40 billion – of which two-thirds is from highly processed, packaged consumer products – and employs one person in five of the manufacturing workforce.

The processed foods sector is inextricably linked with agriculture and is a valuable ally to Australian farmers and producers. Around 90 per cent of its ingredients are sourced domestically, providing a solid link in the value-adding chain and contributing substantially to the country's overall economic position.

Agricultural raw commodities account for as much as a third of the cost of highly processed or transformed products. The future of the processed foods industry may provide a valuable pointer to the future of agriculture in general.

The industry accounts for some 12 per cent of national Gross Domestic Product (GDP), eight per cent of employment, and 23 per cent of the nation's exports.

Australia is well placed to take a leading role in international agri-food markets, with a range of comparative advantages that allow potential for immediate reward and future development.

Major assets include the high quality of safe, uncontaminated foodstuffs, which has given exporters a strong reputation overseas.

Geographically, Australia is ideally situated to take advantage of markets throughout the Asia–Pacific region, both for domestic suppliers and international investors. The relatively temperate climate produces an abundance of natural raw materials for the food processing industry.

The agri-food industry is also setting benchmarks to place itself at the forefront of world's-best practice and to exploit these inherent and created advantages.

Above: An assortment of processed Australian grain products. With an abundance of high-quality primary produce readily available, the processed foods and beverages industry is Australia's largest manufacturing sector. Value adding is the key to improved export revenues. Processed foods are crucial to Australia's standing as an exporting nation, earning the country about $11 billion per year and accounting for some 15 per cent of total exports

A FUNDAMENTAL SHIFT IN EMPHASIS

The Australian food industry is undergoing an unprecedented transition – from an industry supplying a fairly static domestic market, to one which is well placed to serve the significant demand for high-quality foods in Asia.

Agri-food industry participants have been encouraged to adopt an outward-looking stance in the past decade.

Traditionally, Australia has been an exporter of bulk commodity, high-volume, low-value agricultural products, such as wheat. These markets are becoming more challenging to access successfully, particularly as major competitors have access to protectionist policies such as subsidies and cheap credit arrangements.

The most profound shift has been in the realisation that retaining as much value as possible within Australia is the way ahead. Targeting processed foods into specific niche markets means not simply a premium on top of the original produce, but a completely new value structure.

Today, companies are increasingly seeing the unparalleled opportunities that exist in marketing their products overseas as well as in the domestic market.

EXPORTING – THE NEW GROWTH INDUSTRY

Australia continues to export more than three times the amount of processed foods and beverages that it imports. At $11 billion in 1996–97, these products made up one-fifth of manufacturing exports.

The catch-cry is value adding – to capture as much of the economic benefit to Australia as possible. This is a major shift in emphasis for producers and manufacturers, who have traditionally shipped bulk items away to be dealt with elsewhere.

Exports of processed food and beverage products have almost doubled in the past decade. Highly processed food exports grew at a much more pronounced rate than minimally processed food, even though the latter still makes up the majority of total processed food exports.

Asia is a major contributor to Australia's agri-food export performance. Since 1996, food exports to Asia have exceeded $10 billion, to which highly processed foods contributed $2.2 billion per year.

From 1988 to 1989, Asia's share of total Australian food trade grew from 48 per cent to a high of 65 per cent, before settling back to 59 per cent in 1996–97.

Highly processed food and beverage export successes include dairy products (milk, cream, cheese, butter and ice-cream); pet food; chocolate and other confectionery; cereal products (including biscuits, breakfast cereals, pasta and bread pre-mixes); margarine; fruit juices; soups; sauces; and wine.

Many of these products are not commonly produced within Asia and reflect a trend away from traditional Asian foods to Western-style foods.

Above: *A baker stacks his shelves with freshly baked loaves*

Opposite: *Manufacturing potato chips in Adelaide, South Australia. Australia exports more than three times the amount of processed foods and beverages that it imports. The advantage of processing our primary produce locally before it is exported is that, by doing so, Australia captures as much of the economic benefit of the transaction as possible*

REPRESENTING THE INDUSTRY

The Australian Food Council (AFC) was established in 1995 to provide a focus and a forum for collective action by the Australian processed foods and beverages industry.

The AFC was formed to give direction and support to the industry, and to help unite individual industry members to set a common agenda and to focus on achievable goals.

The majority of the industry's main participants contribute to the AFC, which aims to encourage a domestic business environment which is conducive to inter-national competitiveness, strong and sustained investment, and business growth and profitability at the value-added end of the food production chain.

SUPERMARKET TO ASIA

As Australia is poised to take advantage of the massive Asian agri-food market, much of the industry's focus has concentrated on that region.

The Australian Government contributed to the push into the Asian region with the formation, in conjunction with industry, of the Supermarket to Asia Council in 1996.

The Council aims to provide strategic direction and leadership in four key areas:

- business competitiveness and productivity;
- sustainability and quality;
- trade and market development; and
- communication.

The Australian Government has appointed a market access facilitator to liaise with companies in the processed foods and beverages industry. The facilitator will consult with companies to identify market access problems and help them identify the impediments and reach their commercial goals.

Opposite and above: *Selections of premium Australian dairy products, including vintage cheddar and blue-vein cheeses and wholesome, full-cream milk. In Victoria, the biggest and lowest-cost dairy-producing State in Australia, a very high proportion of milk output is 'manufacturing milk' – milk processed into products such as cheeses, butter and skim milk powder. The dairy industry is one of the country's most dynamic industries. Over the past five years, it has attracted an estimated $1 billion in new capital investment and exports have grown to represent about 40 per cent of production, or nearly $2 billion*

Other advances include providing direction to industry participants in terms of computer and 'electronic commerce' technology, assisting with a user-friendly market access guide and database, and generally streamlining market entry for expanding export businesses.

One initiative of the Council is the 'Quality Food Australian Marketing and Promotion Program', which aims to develop markets by bringing agri-food exporters together in Asian markets, and to achieve a more effective recognition of the quality of Australian products in these markets.

The Council's efforts to improve the efficiency of Australia's agri-food industry include addressing vital taxation reform issues, establishing a gene technology regulatory framework, food regulation reviews, national safe food systems and a review of food hygiene standards.

Opposite: *A group of men in hygienic uniforms manufacturing sausages*

Above: *An assortment of processed and fresh meats displayed for sale in a supermarket's delicatessen section. Major export markets for Australian meat products include Japan, the United States, Korea and the Middle Eastern countries. Australia's meat producers are at the forefront of industries taking a market-driven approach to production. Many companies are successfully exporting meat from animals that have been fed special diets to improve the attractiveness of their taste and texture to discerning consumers in the overseas markets*

COMPARATIVE ADVANTAGES

The processed foods industry has a series of comparative advantages which place Australian exporters in the box seat.

Paramount among these is a reputation for quality, hygiene, safety and freedom from contaminants. The safety of food is a totally non-negotiable criterion – and producers and legislators go to great lengths to ensure that this area is not compromised.

Australia has an abundance of natural resources, and a relatively benign climate that provides a readily accessible, generally cost-competitive supply of raw commodities.

Australians have access to leading-edge technology, world-class manufacturing expertise and skilled, flexible management and marketing. Australia's research and development expertise is recognised throughout the world, with a strong food technology base, including twelve Co-operative Research Centres in critical areas of food research.

The high participation of government in research and development in terms of Gross Domestic Product expenditure means that targeted fundamental research can be achieved. In this environment, Australia is participating in programs which are not immediately commercial, but provide a framework for future groundbreaking studies.

Government and industry both contribute to programs aimed at increasing market penetration and effectiveness, which can be commercialised rapidly, along with strategic scientific collaborations within the Asia–Pacific region.

Australia's multicultural society, with its demand for a diversity of food products, prepares the local industry to meet the varied demands of export markets.

LOOKING TO THE FUTURE

Success in agribusiness has undoubtedly been made more challenging by regional economic fluctuations and an increasingly competitive global marketplace.

However, Asia still presents major export and investment business development opportunities for Australian companies because of its proximity and because of continuing growth in the medium to long term.

The projected opportunities are next door in the most populous food markets of the world, particularly in the Asia–Pacific region, and it will be a matter of exploiting that potential with sound, pro-active planning and development.

Australia will also need to form strategic alliances and recognise the potential of integration into the broader regional agribusiness industry.

Maximal advantage will be derived from technical collaboration only if it serves as a bridge for Asian investment in Australian industry, and for Australian investment in

BORIS

the industries which develop in the Asia–Pacific region with the benefit of these technologies.

Exporters are responding to the fact that only about ten per cent of total food demand in Asia is for Western foods, and are tuning their production to traditional foodstuffs. This is particularly relevant in leaner economic times, when the middle classes tighten their belts and among the first things to be crossed off the list are the more 'exotic' imports.

GLOBALISATION AND LIBERALISATION

Competition for Asian market-share is intensifying, with substantial intra-Asian trade and increasingly potent competition from new entrants such as South America and South Africa.

While the Asia–Pacific region has been earmarked as a major expanding food market, it is also predicted that much of the demand will be satisfied from within. Estimates predict that up to 90 per cent of increased demand will be internally provided.

However, given the size of the market, there are still countless opportunities for exporters who are prepared to isolate their niche markets and pursue them strategically and intelligently. Appreciating the differences and working within the different systems is vital. Those companies who expect the familiarity of Western-style trade laws and power over competition will be adding unnecessary barriers to their own success.

On the domestic scene, Australian legislators will aim to foster an environment where local companies are able to trade on a more even footing with their international competitors.

Local and international trade liberalisation will help to create more opportunities offshore for Australian exporters to find niche markets. However, manufacturers will have to be aware of some of the threats that this will provide domestically, as competition increases on the home front.

Ben Calvert

Above: A mushroom, garlic and pepper prepared pasta meal produced by Unilever, a leading Australian food processing company. Exports of Australian processed foods to Asia include both Western-style and traditional foods. The Asia–Pacific region is widely regarded as an important expanding food market. However, demand for Australian processed foods is also increasing domestically

Ricegrowers' Co-operative Limited

Ricegrowers' Co-operative Limited is Australia's most successful producer co-operative and the last global free trader of rice. Located in the Riverina region of New South Wales (NSW) – one of the few rice-growing regions left with none of the world's serious rice pests or diseases – the Co-operative exports around 85 per cent of its annual rice crop of approximately 1.2 million tonnes to more than 40 countries.

The health of the rice-growing area is achieved through a combination of strict government control of rice-based materials entering Australia, total quarantine of the Riverina region, geographical isolation, extremes in summer and winter temperatures and the unique rotation system.

Under this system, rice is only grown on one-third of a property during a season. A natural rotation of land use following rice-growing may include nitrogen-enriching pastures with sheep grazing, cereal crops or field resting. This maintains the health of the soil and helps control weeds and pests, greatly reducing the need for artificial fertilisers and chemicals.

The NSW rice industry is the world's most efficient producer of rice, with the highest average commercial rice yields. It is also the lowest user of artificial fertilisers and chemicals among the world's rice producing countries, producing both the long-grain rice favoured by many Asian countries, and the medium-grain rice favoured by Korea and Japan.

Ricegrowers' Co-operative Limited uses the most advanced technology to ensure that rice is grown, harvested and milled to the highest standard.

Once harvested, the rice is milled in one of six modern computerised rice-milling plants operated by the Co-operative. About 800 brand name and pack size configurations are produced for Australian and international markets, including Sunwhite Calrose and Sunlong, both softer rices that carry colours and flavours well.

The Co-operative continually works to meet consumer needs by developing new products and uses for rice. Research plays an important role. An eight-year study of rice bran, for example, yielded a world-first technique for neutralising the enzyme which previously caused milled rice bran to be unfit for human consumption.

Top: *The rice industry has led the drive for balance between the agricultural potential of land and long-term environmental stability*

Middle: *The Co-operative takes advantage of 'counter-season' benefits by marketing fresh, quality rice to niche markets in the northern hemisphere from harvest in early February each year*

Bottom left: *The Co-operative conducts a highly integrated logistics program to ensure that customer specifications are satisfied by the quality products supplied and that food safety is a number-one priority*

Bottom right: *High in carbohydrates, rice is a valuable source of energy for healthy living*

Ricegrowers' Co-operative Limited has instigated a Research and Technology Commercialisation Centre to research developments of rice-based foods for future domestic and international markets.

Please refer to page 199 for contact details

Roast vegetable pasta salad

Unilever Foods

Unilever Foods, a division of Unilever Australia, is an integral part of the massive Unilever international network, devoted to the pursuit of Unilever's corporate purpose: 'meeting the everyday needs of people everywhere'. Unilever Food's goal of sustainable growth and development is tied to its pursuit of ever-increasing economic and social progress for its customer bases.

The international parent, Unilever PLC, is one of the world's largest and most successful consumer goods companies. It is estimated Unilever services over half of world consumer demand in some form.

Unilever is listed on the London, Rotterdam and New York stock exchanges; the organisation held 35th position on the Fortune Global 500 company rankings in December 1997. Fortune also ranked Unilever as the tenth most profitable multinational company.

Operating globally in 88 nations with a workforce of more than 270,000, Unilever has the financial backing and experience to pursue premium market opportunities and maintain expansion, support, research and development, in the eleven regional operations in South-East Asia and Australia.

Unilever's organisational structure ensures that its operating companies meet local needs and market preferences, simultaneously deriving maximum benefit from international co-ordination of activities including research and development. Unilever is a true multi-local, multinational organisation.

The benefits of Unilever Food's international network are substantial. Unilever Foods produces well over 1,000 food items, from tea beverages to sauces, soups, stocks, canned fish, margarines, flavoured rice, noodles and ice-cream to name but a few. Although cultural differences naturally dictate regional variations of taste, many products are universal. Although Australian-based, some of Unilever Foods' many brands are well-known throughout the world, including premium trademarks like Magnum ice-cream, Vienetta dessert, Lipton tea and Flora margarine.

As well, Unilever Foods accounts for an abundance of famous Australian consumer brands, although most shoppers wouldn't be aware of the Unilever name behind the product. Influential Australian product ranges include Continental soups, stocks, sauces and noodle snacks, John West canned fish and produce, Rosella sauces and soups, Streets ice-cream, Deb instant mashed potato, Flora and Becel margarines,

Bushells and Lipton teas, Raguletto pasta and gourmet sauces, Ally canned seafood and Sakims Asian sauces. Australasian turnover for 1997 was over one billion Australian dollars.

Uniquely placed in Australia, between booming Asia and the Pacific rim, Unilever Foods delivers optimum efficiencies with fast and well-developed access to premium quality produce from Australia's most productive farming regions. Unilever Foods strives to be the first choice in its respective markets, ensuring products and processes comply with stringent in-house quality, safety and legal requirements. Often Unilever's own standards exceed those of international governments.

High standards of quality control are employed throughout the entire supply chain, encompassing harvesting, pre-processing, production, packaging, storage, distribution, sales, usage and customer feedback. To ensure product safety and quality, Hazard Analysis Critical Control Point (HACCP) is used to systematically control manufacturing processes. Contamination sources are therefore contained.

The world's favourite tea brand, blended with teas from many countries

Above: *Quality favourites*

Opposite: *Great ideas for great meals*

Sustainability of suppliers' agricultural practices is of utmost importance to Unilever, not only for continuity of supply but also to ensure environmental responsibility. Unilever's global strategy for managing environmental impact includes conservation of resources, energy efficiency, product stewardship, integration of environmental issues into business management, continuous improvement, attention to the entire supply chain, and waste management and minimisation.

Research has indicated raw material production and end-user wastage are the major contributors affecting the environment, rather than actual manufacturing processes. Consequently, while all areas are monitored and changed accordingly, extensive attention is directed toward improving particular problem fields. To this end, Unilever Foods in Australia actively pursues reduction of waste, conservation of energy, and exploration of opportunities to re-use and recycle wherever possible.

Unilever Foods actively works with growers to maximise usage of drip irrigation to minimise vital water-resource wastage. Results show crop yields are substantially enhanced with parasite numbers drastically reduced, meaning less need for chemical fertilisers and pest controls.

An example of Unilever Foods' resource conservation initiatives is its Tatura production facility, situated in the Goulburn Valley of the southern Australian State of Victoria. It is one of Unilever's four largest international tomato processing centres. The three others are situated in the United States, Brazil and India, and through implementing a scheme modelled on Unilever's Merced, California, tomato processing plant, a large proportion of the nutrient-rich wastewater at Tatura is now recycled as an irrigation and fertiliser resource for property located near the plant.

Overall, brand categories are supported by quality assurance and technology centres and are also enhanced by two regional Innovation Centres in Sydney, covering culinary and beverage products. Studies have shown Sydney to be an ideal test region as people from nearly every Asian and European background live in the city and surrounds. Consequently, an ever-broadening demography of industrial and end-user consumers reap benefits into the future.

Australia's products have a reputation for premium choice, quality and accessibility. The uncrowded and unpolluted land mass allows extensive areas to be utilised for maximum benefit in producing only the highest quality of products. This is the 'Australian advantage': a source of premium products within an environmentally friendly and sustainable framework. Australia's ability to compete with the world's economic leaders, along with its proximity to Asian and Pacific markets, ensures the country's economic and environmental viability is seen as long term.

Unilever Foods realises a competitive edge is paramount to success, exporting an array of processed foods to the Asia–Pacific region. Demonstrating extensive coverage of its Australasian base, Unilever Foods is also well-positioned for competitive opportunities well into the future.

Please refer to page 203 for contact details

BG·7741

CHAPTER THIRTEEN

Oils and Specialty Foods

PRISTINE CONDITIONS FOR NATURAL OILS

Australia's clean waters, rich soils and abundance of sunlight and space are key factors in its capacity to produce the full range of plant oils used for cooking, fragrances and health care.

In the 1960s, farmers took up the challenge of meeting worldwide growth in demand for oilseed products, while processors have continued to invest heavily in leading-edge technology that ensures optimum quality in the finished product.

New investment from the private and public sectors into both consolidated and emerging oilseed products, as well as into essential oils, are ensuring that Australia remains poised to meet new demand from abroad.

At the same time, some of the continent's oldest foods are now being rediscovered for their unique and exciting flavours, creating a vibrant and rapidly expanding industry that represents something of Australia's unique and diverse character.

CUTTING SATURATED FATS

Oilseed production in Australia was comparatively small until the 1960s, when consumers around the world began recognising the health benefits of using vegetable oils to replace butter and other saturated fats.

Since then, Australia has progressively increased its oilseed production to over 1.5 million tonnes per year, and is now regarded as a world leader in research and development into oils and fats.

Australian oilseed processors have also invested heavily in upgrading and expanding their refining and crushing capacity. According to the Australian Oilseeds Federation, this investment has exceeded $80 million in the past five years, leaving the industry well positioned to capitalise on its natural advantages.

Above: *Straws of assorted oils showing a variety of hues and clarity. Since the 1960s, when consumers around the world started recognising the health benefits of using vegetable oils to replace animal fats, Australia has progressively increased its oilseed production. It is now regarded as a world leader in research and development in this area*

CANOLA IS THE STAR

Soft vegetable oils such as canola, sunflower, soybean and cottonseed are the most prominent of those produced in Australia, with canola proving itself as the major success story of the industry, experiencing a 400 per cent increase in production in recent years.

Canola is a high-quality, mainly mono-unsaturated oil which is used in margarine, cooking oil and salad dressings, and is regarded as being superior to polyunsaturated oils as a means of minimising the threat of cholesterol and heart disease.

While Australia's output of olive oil remains small compared to traditional Mediterranean producers, the industry is expanding rapidly as worldwide demand continues to rise.

With a vast portion of the country particularly suitable for olives, it has been estimated that the number of trees in full production by the new millennium could be as high as 1.5 million.

Customers and consumers of Australian oils can also expect the industry to significantly boost its output capacity of other common types, particularly canola, soybean and sunflower, together with other new or emerging oils like sesame and mustard.

The foundations of this growth will again be Australia's climatic and environmental suitability, together with the adoption of sophisticated agricultural and processing technologies.

ESSENTIAL OILS

Like the vegetable oil industry, the Australian essential oils industry has also been experiencing rapid expansion in recent years and now has a current estimated value (excluding tea-tree oil) of about $6–7 million at the farm level and $20–25 million wholesale/retail per year.

According to the Rural Industries Research and Development Corporation, essential oils are largely used in the food-flavouring and fragrance industries, although they are becoming increasingly important in aromatherapy and other health areas.

Essential oils currently being produced in Australia on a commercial scale include boronia, blackcurrant bud, citrus, clary sage, dill, eucalyptus, fennel, lavender, parsley, peppermint and spearmint.

On top of this, Australia remains the world leader in the production of tea-tree oil, which is marketed as a natural antiseptic and antifungal agent and is used in a wide range of health care products.

Since the establishment of plantations in the mid-1980s, production of tea tree oil has increased from about 12 tonnes per year to more than 350 tonnes.

Previous page: *A field of rape, or canola, in full bloom, with its characteristically bright yellow flowers dominating the landscape. Canola is a variable herb,* brassica napus, *which is widely cultivated as a fodder plant and for its seeds, which yield rapeseed (canola) oil. Canola is a high-quality, mainly mono-unsaturated oil, regarded as having health benefits superior to polyunsaturated oils*

Top: *Two handfuls of Macadamia nuts. A small tree native to eastern Australia, the hand-shelled nuts of the macadamia are probably the best-known of the many varieties of native Australian nuts, seeds and vegetables leading the growth of the bush foods industry*

Bottom: *Olives hanging on the tree in a South Australian olive grove. Popularised in many Mediterranean dishes, olive oil is sought-after for its health-giving properties and distinctive taste.*

Opposite: Boronia pulchella, *photographed in Canberra. An Australian shrub, usually with rose-pink or brown flowers and a strong perfume, the boronia yields one of the leading essential oils commercially cultivated in Australia*

And because many recently established plantations are yet to come into production, it is expected that production levels will continue to rise well into the future.

As the natural home of the eucalypt, Australia has also been producing and trading eucalyptus oil, known around the world for its suitability for use in pharmaceuticals, cleaning products and flavourings, for more than 140 years.

Research and development into essential oils is now aiming to capitalise on Australia's advantage as a reliable supplier of high-standard oils produced and extracted with advanced technology, in order to ensure that the industry is able to meet future demand from overseas.

BUSH FOODS

As appreciation of Australia's indigenous cultures has continued to grow, so too has awareness of the country's native ingredients.

Australia's commercial bush food industry now represents a significant contribution to the diversity of world food, with more than three dozen varieties of fruits, nuts, seeds and vegetables being readily available.

Bush foods are increasingly used by chefs as raw produce or to bring new spice to old recipes, as they possess unique flavours which are unavailable in other products.

Some of the most frequently sought varieties are bush tomatoes, Davidson's plum, lemon aspen, lemon myrtle, mountain pepper, muntries, riberries, quandong, warrigal greens and wattleseed.

Often harvested from the wild, but increasingly cultivated in horticultural enterprises, bush foods are generally sold as raw produce to value-adding native food processors, gourmet food or cottage-industry processors, restaurants and caterers which use local native foods.

Bush foods are increasingly used by chefs as raw produce or to bring new spice to old recipes, as they possess unique flavours which are unavailable in other products

Above: A recipe by chef Andrew Fielke using native Australian yabbie, accompanied by whiting tail – ingredients supplied by Bush Tucker Supply Australia. Native Australian ingredients are increasingly sought-after to bring new and exciting flavours to established cuisines

The bush foods industry has experienced steady growth since the mid-1980s. In 1995–96, the farmgate equivalent of its gross value was estimated at $10–12 million.

As the appreciation of indigenous foods grows, and as this young but burgeoning industry develops new ways of promoting and marketing its produce, consumers in search of organically produced food will have access to an ever-growing list of bush produce.

GROWING INTO THE FUTURE

The oilseed and essential oil industries have played an integral role in the Australian agricultural sector's unprecedented push into new foreign markets in the past decade.

And with ongoing investment from processors, together with extensive research and development, Australia is well positioned to keep up with future demand from South-East Asian countries and other markets.

Australia's advantage as a supplier of clean, high-quality produce, so apparent in its exports of crops, fruit, vegetables, meat and wool, is the other factor ensuring its long-term future competitiveness in the supply of oilseed products, essential oils and specialty foods.

Paul Sellars

Gardner Smith

Over more than 75 years, Gardner Smith has built a strong reputation as a reliable supplier of agricultural commodities. It is also developing into a world force in the processing, trading, storing and distribution of vegetable oils, animal fats and natural essential oils.

The company's principal focus is on vegetable oils such as canola, palm, soybean, sunflower, safflower, linseed and coconut oil, together with tallow and protein meal. Its products, being natural, are both renewable and biodegradable.

These are used by its customers to produce items essential to everyday life – foods, flavours, soaps, toothpaste, cosmetics, toiletries, perfumes, plastics, lubricants, rubber products and animal food.

Diversification from the original tallow business began in the 1970s, and the company quickly became a leading supplier of palm, lauric and soft oils to industrial consumers. During the 1990s, Gardner Smith has grown and diversified further, with increased trading activity and vertical expansion into storage facilities, tallow and oil processing, and shipping operations.

From bases in Australia, Singapore, New Zealand, China, Africa, Europe and the United States, its activities now straddle the globe, processing and sourcing commodities, and serving customers on all continents.

Essential to Gardner Smith's service to customers is its people strength, and the ability to maintain a reliable supply of consistent, quality products.

Terminals for bulk liquid handling are integral to Gardner Smith's operational bases, to store its own products and its customers' non-hazardous bulk liquids. The company offers heated and insulated storage, drumming, road and sea transport and the convenience of shoreside locations.

Gardner Smith continues to seek new horizons – be it new products, new markets or new applications – based on its people strength, customer focus, supply chain management and its growing research and development program.

Top to bottom: *A canola field; lavender bloom in January; Pacific terminals – Brisbane facility; Gardner Smith is developing into a world force in the processing, trading, storing and distribution of oils*

Please refer to page 191 for contact details

Australian Food Merchants

Over 30 years in business, Australian Food Merchants (AFM) of Darwin has grown from a small fruit and vegetable providore to a quality gourmet food supplier exporting into Asia and the United States.

Where once AFM involved itself simply in locally grown, everyday table items, today it supplies specialties such as Australian native ingredients and fresh reef fish from north Australian waters.

The company successfully exports hydroponically grown fancy lettuce, fresh herbs, red capsicums, gourmet salad mixes, smoked salmon, lemon myrtle, native bush tomatoes and cheeses from the famous King Island range.

By sourcing its range from all over Australia, AFM is able to supply reliable quantities of fresh produce all year round.

And as well as providing table items, AFM can supply talented chefs and promotional material to its offshore clients to demonstrate the almost unlimited potential of Australian foods.

Above: *Australian Food Merchants supplies quality food to Asia and the United States, including hydroponically grown lettuce and fresh reef fish*

Please refer to page 185 for contact details

Chateau Yaldara

GOLDEN GRAPE SPREAD

A wholesome new food which contains no cholesterol, preservatives, artificial colourings or flavours is being produced from the world-famous, sun-laden vineyards of South Australia.

Golden Grape Spread is a pure, versatile and nutritious gourmet health spread made by Chateau Yaldara – one of the most famous wineries in the premium Barossa Valley wine region.

The result of more than five years of research and development, it represents an international breakthrough in food technology and an exciting step in humankind's long association with the treasured grapevine.

Golden Grape Spread contains five times the vitamins, minerals and trace elements of equivalent spreads such as honey. Because its sugars are absorbed immediately into the bloodstream, it is a source of instant energy which is proving popular with adults and children alike, particularly sportspeople – to say nothing of the recently discovered slimming properties of glucose.

Low-cost and highly versatile, the spread is available in natural, carob and chocolate flavours, and has a shelf life of up to 10 years.

Above: *Chateau Yaldara's gourmet Golden Grape Spread*

Please refer to page 189 for contact details

Olivegrove Trading Company

GOOD OIL FROM A GREAT REGION

Olivegrove Trading Company, at Robinvale on the Victoria/New South Wales border, is located in a perfect area for growing olive trees.

Involved in horticulture for over 35 years, the company now concentrates on olive oil production, marketing the product under the brand name Robinvale Estate Olive Oil.

At present, the company harvests table olives from a 60-hectare grove of Verdale and Manzanilla trees. However, it is establishing 55 hectares of olive trees specifically for oil production.

The 1998 harvest resulted in 8,000 litres of high-quality virgin and extra virgin oil. The company both produces and markets its olive oils, forecasting that production may reach 500,000 litres in the future.

Top chefs in Australia are increasingly using Robinvale Estate Olive Oils because of their quality, flavour and freshness. The olive oils are available direct from the company, at specialised delicatessens and in many gourmet outlets around the country.

Right: *Robinvale Estate quality olive oil, renowned for its flavour and freshness*

Please refer to page 196 for contact details

Agriculture, Fisheries and Forestry – Australia (AFFA)

GPO Box 858
Canberra ACT 2601
Edmund Barton Building
Blackall Street
Barton ACT 2600
Internet: http://www.affa.gov.au
Phone: +61 2 6272 3933
Fax: +61 2 6272 5161
Ken Matthews, Secretary
Bernard Wonder, Executive Director, Corporate Group
See pages 28–31

Ardmona Fresh Fruit Packers Pty Ltd

ACN 005 723 240
Corner Turnbull and Lenne Roads
Ardmona VIC 3629
Email: VARA@Diesel.net.au
Phone: +61 3 5829 0083
Fax: +61 3 5829 0043
Santo Varapodio, Managing Director
Rocky Varapodio, General Manager
See page 138

Asparagus Growers of Sunraysia Inc

PO Box 987CP
Mildura VIC 3501
Phone: +61 3 5024 5876
Fax: +61 3 5024 6585
Internet: www.users.bigpond.com/asparagusgrowers
Email: AsparagusGrowers@bigpond.com
Peter Midgley, President
Val Nichols, Secretary
See page 133

Australian Food Merchants

ACN 009 615 972
PO Box 39
Parap NT 0804
Phone: +61 8 8981 8597
Fax: +61 8 8981 9408
Michael Dinoris, Director
Neville Pantazis, Director
See page 182

Australian Pure Fruits Pty Ltd

ACN 063 269 812
PO Box 465
Lobethal SA 5241
23 Gumeracha Road
Lobethal SA 5241
Email: business@purefruits.com.au
Internet: http://www.purefruits.com.au
Phone: +61 8 8389 6966
Fax: +61 8 8389 5012
Peter Winkley, Managing Director
Peter McDougall, Company Secretary
See page 133

AWB Limited

ACN 081 890 459
GPO Box 4562
Melbourne VIC 3001
528 Lonsdale Street
Melbourne VIC 3000
Email: mrogers@awb.com.au
Internet: www.awb.com.au
Phone: +61 3 9209 2010
Fax: +61 3 9602 3436
Trevor Flügge, Chairman
W Murray Rogers, AM, Chief Executive Officer
NSW
GPO Box 4081
Sydney NSW 2001
WA
PO Box 542
Cloverdale WA 6105
See page 118

Berri Ltd

ACN 008 077 889
60 Jolimont Street
East Melbourne VIC 3002
Phone: +61 3 9654 7577
Fax: +61 3 9658 8127
Doug Shears, Chairman
John Cook, Managing Director
See page 155

Bundaberg Brewed Drinks Pty Ltd

ACN 009 740 863
147 Bargara Road
Bundaberg QLD 4670
Email: gingerbeer@bundaberg-brew.com.au
Internet: bundaberg-brew.com.au
Phone: +61 7 4152 5522
Fax: +61 7 4153 1004
Cliff Fleming, Managing Director
Alan Baumbartner, Research and Development Director
ehl House
Export and Sales Office
3 Chermside Street
Newstead QLD 4006
Phone: +61 7 3252 9760
Fax: +61 7 3252 9767
See page 156

Business Ballarat

PO Box 655
Ballarat VIC 3353
City of Ballarat
Town Hall, Sturt Street
Ballarat VIC 3350
Email: bedb@ballarat.vic.gov.au
Internet: www.ballarat.vic.gov.au
Phone: +61 3 5320 5620
Fax: +61 3 5333 4117
David Miller, Executive Director
See page 57

Cargill Australia Limited

ACN 004 684 173
GPO Box 58A
Melbourne VIC 3001
Level 12, 432 St Kilda Road
Melbourne VIC 3004
Phone: +61 3 9268 7200
Fax: +61 3 9867 8774
Ralph Selwood, Managing Director
Mike Konowalous, Financial Controller
Cargill Foods Australia
Dampier Street
Wagga Wagga NSW 2650
Phone: +61 2 6938 3000
Fax: +61 2 6938 3092
Cargill Salt
Level 12, 432 St Kilda Road
Melbourne VIC 3004
Phone: +61 3 9268 7200
Fax: +61 3 9867 8774
Cargill Oilseed
Level 12, 432 St Kilda Road
Melbourne VIC 3004
Phone: +61 3 9268 7200
Fax: +61 3 9867 8774
Cargill Grain
Level 12, 432 St Kilda Road
Melbourne VIC 3004
Phone: +61 3 9268 7200
Fax: +61 3 9867 8774
See pages 68–69

Chateau Yaldara Pty Ltd

ACN 007 654 462
PO Box 352
Lyndoch SA 5351
Email: gbc@yaldara.com.au
Internet: www.yaldara.com.au
Phone: +61 8 8524 4200
Fax: +61 8 8524 4112
Robert Thumm, Director and Senior Wine Maker
See page 182

Columbus Line Australia Pty Ltd

ACN 000 807 858
GPO Box 5340
Sydney NSW 2001
Level 3, 150 William Street
East Sydney NSW 2011
Email: adresche@hsdgsyd1.columbusline.com
Internet: www.columbusline.com
Phone: +61 2 9373 1000
Fax: +61 2 9373 1100
Achim Drescher, Managing Director
Dr Andreas Opatz, Financial Controller
VIC
Level 15, 607 Bourke Street
Melbourne VIC 3000
QLD
Level 13, 500 Queen Street
Brisbane QLD 4000
SA
61 Hindmarsh Square
Adelaide SA 5000
WA
6 Short Street
Fremantle WA 6160
See pages 86–87

Costa's Pty Ltd

ACN 004 796 918
PO Box 303
Geelong VIC 3220
2 Myers Street
Geelong VIC 3220
Email: exec@costas.com.au
Internet: www.costas.com.au
Phone: +61 3 5222 1122
Fax: +61 3 5222 1112
Frank Costa, Managing Director
Robert Costa, Financial Director
See pages 134–135

Darling Downs Bacon Co-operative Association Ltd

ARBN 051 834 309
PO Box 157
Toowoomba QLD 4350
Email: tomsay@ddbacon.com.au
Internet:
www.businessaustralia.com.au/listing.phtml/ddbacon
Phone: 1300 366 744
Fax: +61 7 4690 6090
Robin Ferris, Chief Executive Officer
David Gifford, Chairman
QLD
231 Riverside Drive
West End QLD 4101
Phone: +61 7 3292 5555
Fax: +61 7 3255 1833
NSW
Unit 9A, 36 Holbeche Road
Arndell Park NSW 2148
Phone: +61 2 9672 1222
Fax: +61 2 9672 1566
VIC
17–19 Pipe Street
Laverton VIC 3026
Phone: +61 3 9369 7666
Fax: +61 3 9369 7078
SA
National Bank Building
Level 1, 75 Dale St
Port Adelaide SA 5015
Phone: +61 8 8240 4353
Fax: +61 8 8240 1309
See pages 49 and 85

edp Packaging and Handling Systems

ACN 005 603 467
PO Box 251
Mooroopna VIC 3629
31–37 O'Brien Street
Mooroopna VIC 3629
Email: edp@shepparton.net.au
Internet: www.edp.com.au
Phone: +61 3 5825 2122
Fax: +61 3 5825 2758
Ian Parsons, Managing Director
John Walton, Company Secretary
See page 139

Gardner Smith Pty Ltd

ACN 000 013 123
Level 14, 61 Lavender Street
Milsons Point NSW 2061
Email: gssyd@gspl.com.au
Phone: +61 2 9959 5919
Fax: +61 2 9959 3347
Otto Hirsch, Chief Executive Officer
Peter Shearman, Director
VIC
46–50 McDonald Street
Numurkah VIC 3636
Phone: +61 3 5862 1666
Fax: +61 3 5862 2391
QLD
105 Tingira Street
Pinkenba QLD 4008
Phone: +61 7 3260 1155
Fax: +61 7 3260 1507
WA
3 Birksgate Road
North Fremantle WA 6159
Phone: +61 8 9335 5999
Fax: +61 8 9336 2872
SA
136–144 Elder Road
Largs Bay SA 5016
Phone: +61 8 8449 2328
Fax: +61 8 8449 2250
See page 181

Gartner Farms

PMB 4
Penola SA 5277
Gartner Road
Glenroy
Phone: +61 8 8736 5019
Fax: +61 8 8736 5006
David Gartner, Director
Greg Gartner, Director
See page 140

Gartner Wines

PO Box 1
Coonawarra SA 5263
Riddoch Highway
Coonawarra SA 5263
Phone: +61 8 8736 5011
Fax: +61 8 8736 5006
Michael Gartner, Managing Director
Philip Gartner, Director
See page 158

Glenmore Export Australia Pty Ltd

ACN 063 401 641
PO Box 222
Glebe NSW 2037
40 Wentworth Park Road
Glebe NSW 2037
Email: fikris@ozemail.com.au
Phone: +61 2 9566 2400
Fax: +61 2 9566 4341
John Sweeney, Managing Director
Fiona Wrigley, Export Manager
See page 88

Goodman Fielder Limited

ACN 000 003 958
Locked Bag 7
Grosvenor Place NSW 1220
Level 42, Grosvenor Place
225 George Street
Sydney NSW 2000
Email: corporate.affairs@goodmanfielder.com.au
Internet: www.goodmanfielder.com.au
Phone: +61 2 9258 4000
Fax: +61 2 9251 5839
David Hearn, Chief Executive
David Clarke, Chairman
See pages 18–23

Harvest Moon

ACN 009 527 913
Leith Road
Forth TAS 7310
Email: Harvest.Moon@tassie.net.au
Phone: +61 3 6428 2505
Fax: +61 3 6428 2952
Neil Armstrong, Managing Director
See page 141

LEADING IN
AGRIBUSINESS

A brand of MaxiTRANS Industries Limited

IHD Pty Ltd

ACN 007 398 003
PO Box 7074
Dandenong VIC 3175
3/47 Princes Highway
Dandenong VIC 3175
Phone: +61 3 9791 3222
Fax: +61 3 9791 3200
Malcolm Roach, Executive Director
Andrew Watson, Business Manager
See pages 32–33

MaxiTRANS Industries Limited

ACN 006 797 173
PO Box 878
Kew VIC 3101
Level 1, 5 Wellington Street
Kew VIC 3101
Email: brees@netspace.net.au
Phone: +61 3 9852 8290
Fax: +61 3 9852 8346
Barry Rees, Managing Director
VIC
31 Hallam South Road
Hallam VIC 3803
QLD
9/11 Platinum Drive
Crestmead QLD 4132
New Zealand
88/90 Takanini School Road
Auckland New Zealand
See page 67

Moraitis Fresh Packaging (NSW) Pty Ltd

ACN 002 306 283
PO Box 428
Sydney Markets NSW 2129
6 Carter Street
Homebush Bay NSW 2127
Email: potatoes@bigpond.com.au
Phone: +61 2 9647 2133
Fax: +61 2 9648 4862
Nicholas Moraitis, Chairman and Chief Executive Officer
Paul Moraitis, Director

Moraitis Fresh Produce (QLD) Pty Ltd
Units 1–9, Block E, Brisbane Market
Sherwood Road
Rocklea QLD 4106
Phone: +61 7 3379 3059
Fax: +61 7 3379 3194

Moraitis Fresh Packaging (QLD) Pty Ltd
Unit 4, 53 McCarthy Road
Salisbury QLD 4107
Phone: +61 7 3275 1550
Fax: +61 7 3275 1228

Moraitis Fresh Produce (VIC) Pty Ltd
Stands 257–259, 'B1' Melbourne Markets
542 Footscray Road
Footscray VIC 3011
Phone: +61 3 9687 1651
Fax: +61 3 9687 3194

Moraitis Fresh Packaging (VIC) Pty Ltd
Corner Ure and Main Roads
Gembrook VIC 3783
Phone: +61 3 5968 1540
Fax: +61 3 5968 1728

See pages 136–137

National Farmers' Federation

PO Box E10
Kingston ACT 2604
14–16 Brisbane Avenue
Barton ACT 2600
Email: nff@nff.org.au
Internet: www.nff.org.au
Phone: +61 2 6273 3855
Fax: +61 2 6273 2331
Ian Donges, President
Dr Wendy Craik, Executive Director

See page 27

Nortas Pty Ltd

ACN 009 555 980
100–104 Mornington Road
Mornington TAS 7018
Email: nortas@nortas.com.au
Phone: +61 3 6244 8099
Fax: +61 3 6244 8102
Richard Doedens, Managing Director
Shane Pritchard, General Manager
See page 109

Olivegrove Trading Co Pty Ltd

ACN 075 064 023
Tol Tol Road
Robinvale VIC 3549
Phone: +61 3 5026 3814
Fax: +61 3 5026 4554
Ralph Gallace, Managing Director
Joseph Antico, Managing Director
See page 183

Orlando Wyndham Group Pty Ltd

ACN 007 870 046
GPO Box 2246
Adelaide SA 5001
33 Exeter Terrace
Devon Park SA 5008
Phone: +61 8 8208 2444
Fax: +61 8 8208 2442
André Boucard, Chief Executive Officer
Stephen Couche, Director, Marketing and Sales
NSW
PO Box 358
Ermington NSW 2115
VIC
PO Box 90
Oakleigh VIC 3166
QLD
39 Dulacca Street
Acacia Ridge QLD 4110
SA
GPO Box 2246
Adelaide SA 5001
WA
PO Box 20
Victoria Park WA 6979
See page 159

P & O Cold Storage

ACN 004 902 998
Corporate Office:
GPO Box 5332
Sydney NSW 2001
Email: john_randall@poaustralia.com
Internet: www.pocold.com
Phone: +61 2 9277 5200
Fax: +61 2 9277 5230
John Randall, Managing Director
Peter Smith, Company Secretary
VIC
90 Coode Road
Footscray VIC 3011
Phone: +61 3 9268 7444
Fax: +61 3 9268 7400
QLD
92 Macarthur Avenue
Hamilton QLD 4007
Phone: +61 7 3308 2400
Fax: +61 7 3268 7954
WA
1 Quarimor Road
Spearwood WA 6163
Phone: +61 8 9418 3300
Fax: +61 8 9418 5224
New Zealand
27 Dalgety Drive
Manurewa
Auckland, New Zealand
Phone: +64 9 268 3939
Fax: +64 9 268 3930
See page 41

Pace Farms Pty Ltd

ACN 003 529 575
Lot D Richmond Road
Dean Park NSW 2761
Email: kpace@pacefarms.com.au
Phone: +61 2 9626 9744
Fax: +61 2 9626 3923
Frank Pace, Managing Director
See page 99

Perfecta Produce

ACN 009 573 648
PO Box 3126
Ulverstone TAS 7315
45 Kimberleys Road
Ulverstone TAS 7315
Email: perfecta@tassie.net.au
Internet: www.perfecta.com.au
Phone: +61 3 6425 6999
Fax: +61 3 6425 5892
Michael Broadby, Managing Director
Gaynor Broadby, Export Sales Director
See page 142

Rangers Valley Cattle Station Pty Ltd

ACN 001 060 402
PO Box 63
Glen Innes NSW 2370
Dundee Road, Rangers Valley
Glen Innes NSW 2370
Email: rvalley@northnet.com.au
Phone: +61 2 6734 4977
Fax: +61 2 6734 4985
Masanao Asai, Managing Director
Malcolm Foster, General Manager
See page 47

Ricegrowers' Co-operative Limited

ARBN 007 481 156
PO Box 561
Leeton NSW 2705
Yanco Avenue
Leeton NSW 2705
Email: webmaster@ricegrowers.com.au
Internet: www.sunrice.com.au
Phone: +61 2 6953 0411
Fax: +61 2 6953 4733
Terry Hogan, Chairman
Richard Day, Chief Executive Officer
See page 169

RIDLEY

Ridley Corporation Limited

ACN 006 708 765
PO Box 493
Epping NSW 1710
Level 1, 39 Oxford Street
Epping NSW 2121
Email: cpriestl@ridley.com.au
Internet: www.ridley.com.au
Phone: +61 2 9877 2121
Fax: +61 2 9877 2100
John Keniry, Chairman and Acting Managing Director
Ed Moloney, President and Chief Executive Officer, Ridley Canada Limited
Ridley AgriProducts Pty Ltd
Level 1, 39 Oxford Street
Epping NSW 2121
Ridley Pet Products
1325 Boundary Road
Wacol QLD 4076
Cheetham Salt Limited
Lowe Street
Corio VIC 3214
Ridley Canada Limited
24th Floor
Richardson Building
1 Lombard Place
Winnipeg
Manitoba R3C OX3
See page 70

ROBERTS ESTATE

Roberts Estate Wines

Head Office
PO Box 19
Irymple VIC 3498
2160–2162 Fifteenth Street
Irymple VIC 3498
Email: rew@irymplepacking.com.au
Phone: +61 3 5024 5704
Fax: +61 3 5024 6550
Winery
83 Game Street
Merbein VIC 3505
James Harrison, Director
David Stevenson, Executive Director
See page 157

STANBROKE PASTORAL COMPANY PTY LTD A.C.N. 008 442 939

DIAMANTINA

Food Victoria

Stanbroke Pastoral Company Pty Ltd

ACN 008 442 939
GPO Box 155
Brisbane QLD 4001
Level 11, 12 Creek Street
Brisbane QLD 4000
Email: stanmd@stanbroke.com.au
Phone: +61 7 3229 1516
Fax: +61 7 3229 6204
Dr Rod Kater, Chairman
John Cox, Managing Director
See page 89

State of Victoria

Food Victoria
Level 13, 55 Collins Street
Melbourne VIC 3000
Email: food@food.vic.gov.au
Internet: www.food.vic.gov.au
Phone: +61 3 9651 9444
Fax: +61 3 9651 9861

Business Victoria
Level 11, 55 Collins Street
Melbourne VIC 3000
Internet: www.business.vic.gov.au
Phone: +61 3 9651 9453
Fax: +61 3 9651 9304

Agribusiness Group: Department of Natural Resources and Environment
Level 15, 8 Nicholson Street
East Melbourne VIC 3002
Phone: +61 3 9637 8498
Internet: www.nre.vic.gov.au

Food Safety Victoria: Department of Human Services
Level 17, 120 Spencer Street
Melbourne VIC 3000
Phone: 1300 364 352
Internet: www.dhs.vic.gov.au

Melbourne Market Authority
Box 1, 542 Footscray Road
Footscray VIC 3011
Email: info@mma.vic.gov.au
Phone: +61 3 9258 6100
Fax: +61 3 9287 7714
See pages 24–26, 40, 48, 56 and 140

Tassal Limited

ACN 009 548 770
GPO Box 1645
Hobart TAS 7001
5 Franklin Wharf
Hobart TAS 7000
Email: tassal@tassal.com.au
Internet: www.tassal.com.au
Phone: +61 3 6211 9611
Fax: +61 3 6224 0373
Richard Watson, Chairman
Phil Cooper, General Manager, Export
Alan Newman, General Manager, Marketing and Sales Australia
See page 109

The Lentil Company Pty Ltd

ACN 059 587 478
PO Box 1145
Horsham VIC 3401
41–43 Roberts Avenue
Horsham VIC 3400
Email: tlcaus@comcirc.com.au
Internet: www.thelentilcompany.com.au
Phone: +61 3 5382 1277
Fax: +61 3 5382 0053
Rob Launder, General Manager
Peter Semmler, Export Marketing Consultant
See page 119

The Original Juice Company

ACN 007 070 313
20 Heaths Court
Mill Park VIC 3082
Email: nick@originaljuice.com.au
Phone: +61 3 9436 6166
Fax: +61 3 9436 6366
Gino Vescio, General Manager
Nick Katopodis, Export Manager
NSW
13–15 Alfred Road
Chipping Norton NSW 2170
Phone: +61 2 9755 3500
Fax: +61 2 9755 3504
QLD
21 Murrajong Road
Springwood QLD 4127
Phone: +61 7 3208 4000
Fax: +61 7 3299 2799
See page 158

Unilever Foods
A division of Unilever Australia Limited

Unilever Foods

ACN 004 050 828
Private Bag 2
Epping NSW 2121
20–22 Cambridge Street
Epping NSW 2121
Email: garry.west@unilever.com
Internet: www.unilever.com
Phone: +61 2 9869 6400
Fax: +61 2 9869 6430
Enzo Allara, National Manager, Australasia
Ian Lawrence, Chief Financial Officer
See pages 170–173

Warrnambool Cheese & Butter Factory Company Limited

ACN 004 032 053
5331 Great Ocean Road
Allansford VIC 3277
Email: WCBF@Ansonic.com.au
Phone: +61 3 5563 2100
Fax: +61 3 5563 2156
John McLean, General Manager
Des McKinnon, Chairman
See page 98

Wirrilla Pty Ltd

ACN 000 179 153
Bundarbo Road
Jugiong NSW 2726
Email: wirrilla@dragnet.com.au
Phone: +61 2 6945 4239
Fax: +61 2 6945 4305
David A Power, Managing Director
Jonathan Harpley, Secretary Business Manager
See page 71

Woolworths Limited

Woolworths Limited

ACN 000 014 675
PO Box Q70
QVB Post Office NSW 2000
Internet: www.woolworths.com.au
Phone: +61 2 9323 1555
Fax: +61 2 9323 1593
John Dahlsen, Chairman
Roger Corbett, Chief Executive Officer
National/State Office Locations:
National Supermarkets Office and NSW State Office
Locked Bag 11, Fairfield NSW 2165
Phone: +61 2 9892 7111
QLD State Office
PO Box 202, Archerfield QLD 4108
Phone: +61 7 3213 4111
SA/NT State Office
PO Box 2339, Regency Park SA 5942
Phone: +61 8 8206 5900
WA State Office
GPO Box D149, Perth WA 6840
Phone: +61 8 9351 5222
VIC State Office
Private Bag 10, Mulgrave North VIC 3170
Phone: +61 3 9263 2444
TAS State Office
20 Lampton Avenue, Derwent Park TAS 7009
Phone: +61 3 6211 6211
See page 143

INTERNET INFORMATION

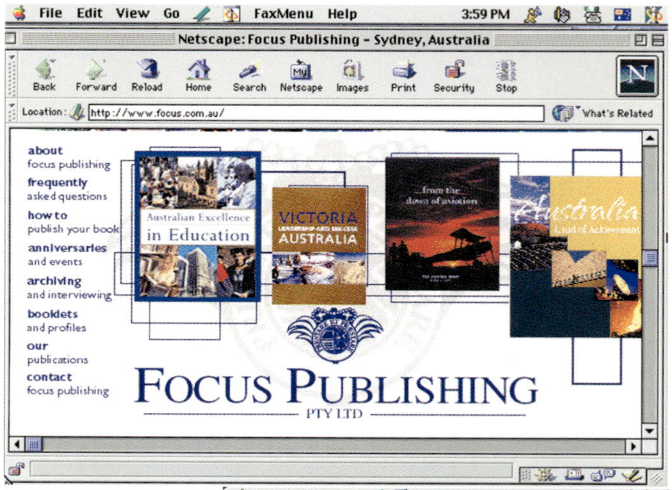

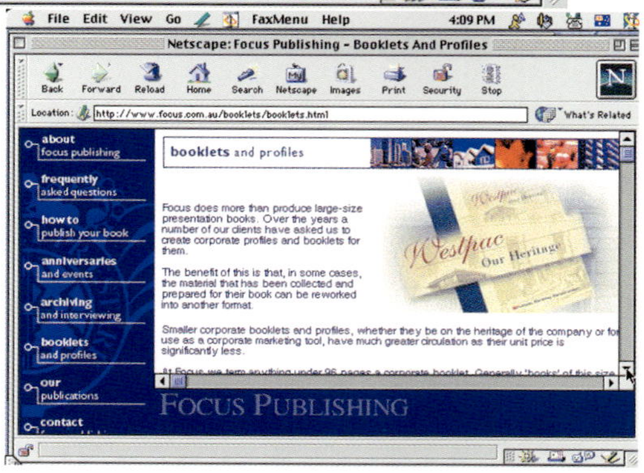

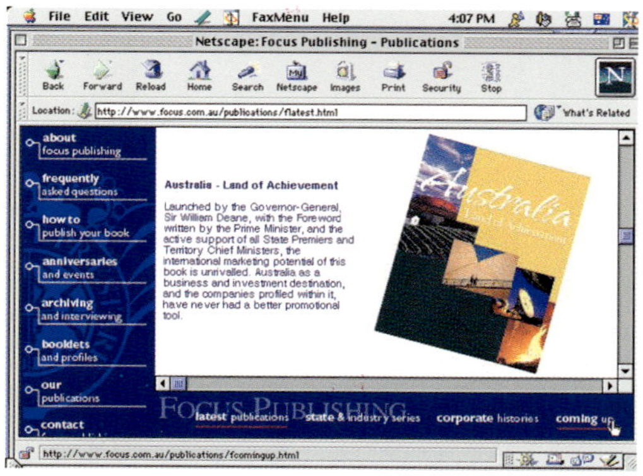

www.focus.com.au

To obtain further information on the organisations participating in this book, simply follow these instructions: Enter the Focus Web site address, www.focus.com.au

- You are now on Focus' home page. Click on the words 'our publications'.
- Scroll down to the bottom of the page, and click on 'latest publications'.
- Scroll down the page to find *Natural Quality: The Produce of Australia.*
- Click on 'Agribusiness Links'.
- You will find a listing of the Web sites of the participating companies and other organisations. These are in light blue type. Click on your chosen institution's address, and it will lead you to their Web page.
- Enjoy!

You may also wish to visit the Focus Web site for information about:

- Focus Publishing
- our other publications
- how to publish your book
- and many more frequently asked questions!

PICTURE CREDITS

Dust jacket: International Photographic Library; p 8: Auscape International/Jean-Marc La Roque; p 9: Tourism Victoria; p 10: BRI Australia Limited; p 11: Stock Photos/Owen Hughes; pp 12, 13: Rosemount Estates Pty Ltd; p 14: (top to bottom) Stock Photos/Pauline Madden; Rangers Valley Cattle Station; p 15: Stock Photos/Otto Rogge; p 16: Stock Photos/Otto Rogge; p 17: Stock Photos/Lance Nelson; p 32: Great Southern Stock; p 33: (top) IHD Pty Ltd; (middle and bottom) Great Southern Stock; p 34: BRI Australia Limited; p 35: Landcare Australia Limited; p 36: Woolworths Limited; p 37: Stock Photos/Peter Fisher; p 38: (top to bottom) Columbus Line/David Moore; University of Technology, Sydney; p 39: Auscape International/Jean-Paul Ferrero; p 42: Stock Photos/Lance Nelson;p 43: Geoff Lung; p 44: (top to bottom) Warrnambool Cheese and Butter Factory; Chisholm Manufacturing; Rangers Valley Cattle Station; p 45: Warrnambool Cheese and Butter Factory; p 46: (top and middle) BRI Australia Limited; (bottom) Meat and Livestock Australia; p 50: Stock Photos/Paul Foley; pp 51–5: Landcare Australia Limited; p 58: Geoff Lung; p 59: Australian Horticultural Corporation; pp 60, 61: Rosemount Estates Pty Ltd/Steven Morris; pp 62, 63: Auscape International/Alby Ziebell; p 64: (top left) University of Melbourne; (top right) Geoff Lung; (bottom) Auscape International/Mike Langford; p 65: (top) Australian Farm Journal; (bottom) Town and Country Farmer; p 66: Inter•Continental Hotels and Resorts Asia–Pacific; p 68: Meat and Livestock Australia; p 69: Australian Picture Library/ZEFA; p 74: Stock Photos/Paul Steel; pp 75, 76: Meat and Livestock Australia; p 77: (top) Woolworths Limited; (bottom left) Meat and Livestock Australia; (bottom right) Malcolm McCosker; pp 78–80: Meat and Livestock Australia; p 81: Tourism Victoria; pp 82, 83: Australian Pork Corporation; p 84: (top) Stock Photos/Kevan Way; (middle and bottom) Australian Poultry Industries Association; pp 90, 91: Australian Dairy Corporation; p 92: (top to bottom) Australian Dairy Corporation/Rick Altman; Stock Photos/Geoff Wharton; p 93: (top left, top right and bottom right) Warrnambool Cheese and Butter Factory; (bottom left) Australian Dairy Corporation; pp 94, 95: Australian Picture Library/Oliver Strewe; p 96: W J's Photographic Services/RIRDC Egg Program; p 97: (top to bottom) Australian Picture Library/John Carnemolla; W J's Photographic Services/RIRDC Egg Program; pp 100, 101: Sydney Fish Market; p 102: Auscape International/Jean-Marc La Roque; p 103: Sydney Fish Market; pp 104, 105: Stock Photos/Tom Stewart; p 106: (top left and top right) Sydney Fish Market; (bottom) Tassal Ltd; p 107: Sydney Fish Market; p 110: Stock Photos/Paul Steel; p 111: Ricegrowers' Co-operative Ltd; p 112: NSW Grains Board; p 113: (top left) NSW Grains Board; (top right) Stock Photos/Bruce Peebles; (bottom left) Stock Photos/ Bill Bachman; (bottom right) NSW Grains Board; p 114: AWB Ltd; p 115: Stock Photos/Lance Nelson; p 116: Stock Photos/Paul Steel; p 117: NSW Grains Board; p 120: Australian Tourism Commission; p 121: Stock Photos/Bruce Peebles; p 122: Auscape International/ C Andrew Henley; p 123: Auscape International/Geoffrey Lea; p 124: (top left) Auscape International/Rob Walls; (top right) Australian Horticultural Corporation; (bottom left) Auscape International/Michel Viard; (bottom right) Australian Horticultural Corporation/Milton Wordley and Associates; p 125: Australian Horticultural Corporation; pp 126, 127: Stock Photos/Otto Rogge; pp 128, 129: Bush Tucker Supply Australia/Vic Cherikoff (www.bushtucker.com.au/); p 130: (top left and top right) Woolworths Limited; (bottom left) Auscape International/ C Andrew Henley; (bottom right) Australian Horticultural Corporation; p 132: Tourism Victoria; p 144: Auscape International/Guy Lamothe; p 145: Australian Dairy Corporation; pp 146, 147: Auscape International/Jean-Paul Ferrero; p 148: (top) Rosemount Estates Pty Ltd; (bottom) Rosemount Estates Pty Ltd/Steven Morris; p 149: Auscape International/Jean-Paul Ferrero; pp 150, 151: Rosemount Estates Pty Ltd/Steven Morris; pp 152, 153: Woolworths Limited; p 154: (top) Australian Horticultural Corporation; (middle) Auscape International; (bottom) Auscape International/Jean-Marc La Roque; p 160: Woolworths Limited; p 161: Unilever Foods; p 162: Woolworths Limited; p 163: Stock Photos/Peter Fisher; pp 164, 165: Australian Dairy Corporation; p 166: Woolworths Limited; p 167: Stock Photos/Tom Cockrem; p 168: Unilever Foods; p 174: Auscape International/Reg Morrison; p 175: Gardner Smith Pty Ltd; pp 176, 177: NSW Grains Board; p 178: (top) Australian Horticultural Corporation; (bottom) Stock Photos/Lance Nelson; p 179: Auscape International/Brett Gregory; p 180: Bush Tucker Supply Australia/Vic Cherikoff (www.bushtucker.com.au/)

All images used in participants' profile pages are supplied by the participant unless otherwise specified.

CONTRIBUTING WRITERS

CATHY BOLT has worked as a senior journalist since 1989 with the *Australian Financial Review*, mostly specialising in farming and agribusiness but including a two-year period as the paper's Melbourne-based industrial relations writer. Previously, she has worked in public relations and for a range of farming publications.

DUNCAN BROWN is a Melbourne-based journalist specialising in rural commodity reporting.

BEN CALVERT has worked in Australia and overseas as a journalist. He is the public relations and marketing director for Gorrie Group Media in Melbourne. Ben specialises in targeted information packages for Australia's major research and development bodies.

DAVID GORRIE is a former journalist, Canberra-based national affairs editor and award-winning photographer. David is the managing director of Gorrie Group Media, a publishing and information management company in Melbourne. His company specialises in information packages and strategies for the food and fibre industries.

JACQUI MACDONALD is a Melbourne-based freelance writer who worked as a journalist for the *Age* for eight years. She has twice been a finalist in Australia's premier journalism awards, the Walkleys, and in 1996 won the Australian Human Rights award for print media.

MALCOLM McCOSKER OAM started writing in November 1961 as a trainee journalist with *Queensland Country Life* and has continued a career as a journalist, editor, author, photographer and rural industries commentator ever since, working in Australia and several overseas countries. He was awarded a Medal of the Order of Australia (OAM) in January 1988 for services to the media, particularly rural journalism.

PAUL SELLARS is the horticultural reporter for the *Weekly Times* newspaper in Melbourne.

TIM TREADGOLD is a Perth journalist who has written for a number of newspapers and magazines over a 30-year career. His current work is seen mainly in *Business Review Weekly*.

TIM STEVENS is a specialist rural journalist, having worked for leading rural newspapers and magazines. This includes stints as the Canberra bureau chief for the *Australian Rural Times* and as a political writer for the *Land*. He has also worked as a national affairs reporter for the ABC's 'Country Hour' program and as Press Secretary to the New South Wales Minister for Agriculture. Most recently, he was the rural writer for the *Australian*, based in the Press Gallery in Canberra. He now runs a vineyard at Mudgee in New South Wales and sells wine under the Abercorn label.

With thanks also to Dallas Alldridge, Dael Climo, John Cokley, Niree Creed, John Hardcastle, Mark Hotston-Jones, Liz Kellaway, Jacqui Macdonald, Karen Prout and Jamie Walls for compiling the company profiles.